The Open University

A Second Level Course

INSTRUMENTATION

Units 3 and 4

Transducers 1
Temperature, Displacement, Force, Torque, Pressure

Prepared by the Course Team

THE OPEN UNIVERSITY PRESS

The Open University Press
Walton Hall, Milton Keynes

First published 1974

Copyright © 1974 The Open University

Designed by the Media Development Group of the Open University.

Printed in Great Britain by
Staples Printing Group
at St Albans

ISBN 0 335 02951 5

This text forms part of an Open University course. The complete list of units in the course appears at the end of this text.

For general availability of supporting material referred to in this text, please write to the Director of Marketing, The Open University, PO Box 81, Milton Keynes, MK7 6AT.

Further information on Open University courses may be obtained from the Admissions Office, The Open University, PO Box 48, Milton Keynes, MK7 6AB.

1.1

D

621.379

OPE

Aims

The broad aim of these two units is to discuss transducers for measuring temperature, displacement, force, pressure and torque.

More specifically, the aims in discussing transducers for each type of *measurand* are:

1   To explain some of the more important physical principles used in transducers for that measurand.

2   To show you some examples of the way that these principles are exploited in proprietary transducers for these measurands.

3   To highlight some of the more important fundamental performance limitations which arise in transducers using the different physical principles.

Transducers can be classified either in terms of their *measurand* or in terms of their *principle of operation*. The aims of these two units can be classified similarly.

The broad aim of Unit 3 is to discuss temperature transducers, with the more specific aims 1, 2 and 3 above.

The broad aim of Unit 4, however, can be classified as either: (a) a discussion of the transducers for the measurands displacement, force, pressure and torque, or (b) a discussion of transducers using the principles of strain gauges, 'pots', magnetic effects, capacitance and piezoelectricity, all of which are used for the measurands discussed in this unit.

It turns out that it is convenient to structure Unit 4 according to the second classification, that is, by principles of operation. Nevertheless, the specific aims 1, 2 and 3 are still applicable.

Objectives

Unit 3

When you have completed this unit you should be able:

1   To describe the thermodynamic (Kelvin) temperature scale and the use of the International Practical Temperature Scale in calibration of temperature transducers.

2   To explain the principles of operation of thermometers using changes in length, volume and pressure.

3   To calculate a temperature measured by a thermocouple, given the temperature of the reference junction, the characteristics of the thermocouple junction and the output voltage of the thermocouple circuit.

4   To discuss the reasons for practical details of some thermocouple examples in terms of their principle of operation and the environment for which they are intended.

5   To calculate the temperature measured by a metallic resistance thermometer or a thermistor, given the resistance and necessary characteristics in either case.

6   To explain the principles of some circuits for measuring resistance and their applicability to temperature measurements made with resistance thermometers.

7 To outline the principle of operation of an electronic thermometer with reference junction temperature compensation.

8 To calculate the radiation from a black body, given its temperature, and to explain how the radiation from a body is modified by its emissivity.

9 To explain the principle of operation of a radiation pyrometer.

Section 6 (Interlude)

10 To explain the reason for the delayed response of a temperature transducer.

11 To identify this response as an exponential and as a feature of a first-order system.

Unit 4

When you have completed this unit you should be able:

1 To describe the elastic sensing elements used in transducers for measuring force, torque and pressure, and which convert these measurands into displacements.

2 To discuss the properties of a proprietary load cell using strain gauges.

3 To explain the principle of operation and limited resolution, and calculate the non-linearity caused by electrical loading, of resistive potential dividers ('pots').

4 To explain the principle of operation of a proprietary pressure transducer using a pot.

5 To explain the principle of operation of a differential transformer and the way that its output voltage represents the direction and magnitude of the displacement.

6 To explain the basic principle of operation of a phase-sensitive detector and how it conditions the output signal from a differential transformer.

7 To explain the principle of operation of a variable inductance (variable reluctance) displacement transducer.

8 To explain the principle of operation and some of the properties of a proprietary inductance-bridge pressure transducer.

9 To explain the three ways in which the capacitance of a capacitor can be varied by a displacement.

10 To show how the output voltage of an a.c. bridge circuit, with capacitors in two of the arms, varies with the values of the capacitors.

11 To show how the combination of a differential capacitive transducer and such a bridge circuit has a linear relationship between displacement and bridge output voltage.

12 To explain the principle of operation of a proprietary differential pressure transducer incorporating a differential capacitor.

13 To describe piezoelectricity and how it is used in piezoelectric force transducers.

14 To analyse the step response of a piezoelectric force transducer and resistive load or, more specifically, of its equivalent CR circuit, and to make conclusions about its low-frequency response.

15 To explain the principle of, and analyse simply the operation of, a charge amplifier, the use of which avoids the variation of sensitivity of a piezoelectric transducer with the capacitance of the connecting cable.

4

Study Guide for Unit 3

Unit 3 is somewhat lighter than Unit 4, so I would suggest that you try to apportion somewhat less than ten hours to it, using any time you are able to save to make an early start on Unit 4. Only one paper in the *Instrumentation* Course Reader* is relevant to Unit 3; this is paper 6, which is quite short. It should only take about fifteen minutes to read. The radio programme 'First-order response' covers the material in Section 6 'Interlude: The response time of first-order transducers'. There will be assignments covering Units 3 and 4 together; allow about three hours altogether for these.

Study guide for Unit 4

The *Instrumentation* Course Reader papers relevant to this unit are papers 3, 4 and 5. Papers 3 and 4 are short and easy to read. Paper 5 takes quite a detailed look at a particular problem of pressure measurement. Read it only if you have time to spare. Allow about one hour for reading these papers, at most. The radio programme 'Magnetic circuits' and the television programme 'Pressure transducers' are associated with this unit.

Introduction

The first task of any electrical instrumentation system is to translate the physical variable or property to be measured (the measurand) into an electrical variable. The devices used to do this are called transducers; they form the subject of Units 3 and 4, as well as of Units 8/9/10.

Before we begin to look at transducers I should like to explain the philosophy behind these two units and what I think you should try to get out of them. There are three basic problems which arise in trying to teach about transducers:

1 The variety of transducers and the technical literature describing them is vast. Virtually any physical effect which causes a change in some electrical property of a material can be used to make a transducer. Two commercially available transducers for measuring the same variable (such as pressure or displacement) may use completely different operating principles.

2 The applications of transducers are even more varied than the choice of transducers.

3 The field is developing rapidly. Improvements to existing types of transducers and the development of new types make judgements about transducers very liable to become out of date. This problem is accentuated by the interval between the time of writing these units (1973) and the time when you will actually be using the information.

As a result of these problems it is not possible for these units to give a comprehensive description of all the transducers you are likely to come across. Similarly, it is not possible to tell you what types of transducers to use in a given application. Almost any real industrial or scientific application has its own special needs and problems, which make such

*Hunter, J. J., and Crecraft, D. I. (1973) Instrumentation, *Open University Press/Holmes McDougall*.

5

generalizations useless, in addition to the problem of such statements becoming dated.

Having said what the units cannot do, I now turn to what they can do, and what you should look for as you read them.

There are a number of physical principles upon which the operation of a wide variety of transducers is based. You should understand these principles, their limitations and their advantages. Then, if you find yourself in the position where you need to choose a transducer, you can find out what transducers are available and, by understanding their operation and the requirements of your particular application, make an intelligent choice.

In the course of these units I shall describe a selection of transducers. The selection is necessarily quite arbitrary and, of course, is very limited. It was chosen to illustrate a wide range of transducer principles and the ways in which they can be applied.

To make the section of the *Instrumentation* course which deals with transducers more manageable, it has been split into two groups: Units 3 and 4 and Units 8/9/10. The particular variables selected for study in Units 3 and 4 are temperature, force, torque, pressure and displacement.

Unit 3 describes transducers for measuring temperature. Unit 4 describes transducers for measuring the other variables. At the end of Unit 3, after the descriptions of temperature transducers, there is a section which I have called an 'interlude'. The interlude discusses the very important topic of response time of transducers. I have placed it after the descriptions of temperature transducers so that I can use these transducers as examples. However, the topic is much more widely applicable than just to temperature transducers. The radio programme 'First-order response' also covers this subject.

The radio programme 'Magnetic circuits' is related to Unit 4. Magnetic circuits are the basis of several types of transducer described in Unit 4. The television programme 'Pressure transducers' describes the construction and operation of two types of pressure transducer discussed in Unit 4.

Unit 3

Contents

1 Temperature scales   9

2 Temperature measurement using changes in length, volume or pressure   12
2.1 Bimetallic strip   12
2.2 Liquid-in-glass thermometer   12
2.3 Sealed-fluid thermometers   13

3 Thermoelectric (thermocouple) temperature sensors   14
3.1 Basic principles   14
3.2 Reference junctions   15
3.3 Transducer example 1: A thermocouple electronic thermometer   19
3.4 Thermocouple materials and construction   19
3.5 Transducer example 2: A range of thermocouples   21

4 Resistive temperature transducers   23
4.1 Basic principles   23
4.2 Construction of metallic resistance thermometers   25
4.3 Transducer example 3: A platinum resistance temperature sensor   26
4.4 Thermistors   28
4.5 Measuring circuits for resistive temperature transducers   29
4.6 Transducer example 4: An electronic thermometer using a thermistor   32

5 Radiation pyrometers   35
5.1 Infrared radiation   36
5.2 Black bodies and their properties   36
5.3 Radiation from objects other than black bodies   38
5.4 Pyrometer optical systems and detectors   41
5.5 Transducer example 5: A radiation pyrometer   42

6 Interlude: The response time of first-order transducers   45

7 Summary   51

Self-assessment answers and comments   53

Section 1

Temperature scales

Measuring temperature is a familiar experience for most people. You have probably used one or two different types of thermometer, such as a mercury-in-glass or an alcohol-in-glass thermometer, or a cooking thermometer using a coiled bimetallic strip (Figures 1 and 2).

Figure 1   A mercury-in-glass thermometer

There are many other ways of measuring temperature, based on a variety of physical principles. In Unit 3 I shall describe transducers for measuring temperature in which changes in temperature are sensed as:

1   changes in volume, length or pressure (used in mercury-in-glass thermometers, bimetallic thermometers, sealed-fluid thermometers, vapour-pressure thermometers);

2   the change in contact potential between different metals (used in thermocouples);

3   the changes in resistance of a metal or a semiconductor (used in metallic resistance thermometers and thermistors);

4   the change in the energy radiated by a hot object (used in radiation pyrometers).

Not only are there many ways of measuring temperature, but there are several scales on which it is measured.

Figure 2   A thermometer using a coiled bimetallic strip

You are probably familiar with at least two scales: the Fahrenheit scale (°F) and the metric scale (which uses °C), called the Celsius (or, less correctly, centigrade) scale. In your foundation course you will have met the temperature scale which is part of the *Système International* (SI). It is a 'thermodynamic' scale in which the unit of temperature is the *kelvin* (K).*   **kelvin**

A proper explanation of a thermodynamic scale is beyond the scope of this course, but, for interest only, I am including a brief description.

A thermodynamic temperature scale is not based on any of the properties mentioned above used in devices for measuring temperature. The scale is established by assigning numerical values for two points. All other temperatures are then defined conceptually from these two in terms of the heat reservoirs of an ideal Carnot engine. One of the points defining the Kelvin scale is 0 K, or *absolute zero*. It is the theoretical minimum   **absolute zero** temperature of any substance. The more you cool a substance, the closer you can bring its temperature to absolute zero, but you can never actually bring the temperature down to absolute zero.

The other defining point on the Kelvin scale is the temperature of what is called the *triple point of water*. If a mixture of ice and water is placed in   **triple point of water** an insulated container so that no heat transfer can take place between the

---

* *Note that the degree symbol (°) is not nowadays used in the Kelvin scale.*

ice–water mixture and its surroundings, the temperature of the mixture and the proportion that is water and the proportion that is ice will remain constant. However, the temperature will depend slightly upon the pressure in the container. Different pressures will give different temperatures. (For example, the temperature at 'standard atmospheric pressure', 101 325 Pa, is the freezing point of water used in the Celsius scale as 0 °C.) There is one particular pressure, 610 Pa, and a corresponding particular temperature, at which some of the water will also be present in the form of water vapour. This particular combination of pressure and temperature at which water, ice and water vapour can coexist is called the 'triple point' of water. Because it occurs at a unique reproducible temperature, that temperature can be used to define a temperature scale. Its value is defined as 273.16 K.

What properties of the temperature scale are set by assigning a value of 273.16 K to the triple point of water?

This value sets the interval, in kelvins, between any two temperatures. It was chosen so that the temperature difference between the boiling point and freezing point of water, used to define the Celsius scale, would be exactly 100 K. In that way, temperatures expressed on the Celsius scale can be converted to the Kelvin scale, simply by adding the Kelvin temperature corresponding to 0 °C. That temperature, the freezing point of water (at a pressure of 101 325 Pa), is 273.15 K. Notice that this is not the temperature of the triple point of water.

The temperature of boiling water at a pressure of one standard atmosphere defines 100 °C. What is its temperature expressed in kelvins?

373.15 K.

A simple way to establish a temperature scale would be to take a mercury-in-glass thermometer, or some other device for sensing temperature, label the points 0 and 100 on the scale by measuring the temperature of freezing and boiling water, respectively (at a pressure of one standard atmosphere), and then divide the scale into 100 equal intervals, extrapolating for higher and lower temperatures.

The trouble with this approach is that the readings given by differently constructed instruments, measuring the same temperature, would differ somewhat at practically all points other than 0 and 100. This would be true even for two mercury-in-glass thermometers made of different types of glass. Moreover, none of the readings on scales such as these would be in perfect agreement with the Celsius scale.

What I am saying is that none of the properties of materials used for temperature sensing vary strictly linearly with temperature. This creates a calibration problem for instrument manufacturers. To enable instruments to be calibrated in terms of the Kelvin or Celsius scales, the International Practical Temperature Scale* (IPTS) has been devised.

In the IPTS, values of temperature are assigned to the eleven reproducible 'fixed points' shown in Table 1. Standard instruments, chosen for their repeatability and stability, with designated interpolation equations, are used at temperatures between the fixed points. In this way the IPTS is made to agree with the Kelvin scale as closely as the current state of measurement technology will allow.

* *The complete text of the IPTS, 1968 version, is published in* Metrologia. *vol. 5 (1969), no. 2, p. 35.*

In order to ensure that the readings of any of the temperature-measuring transducers described in the following sections are on the Kelvin scale (or on the Celsius or Fahrenheit scales, since these are defined in terms of the Kelvin scale), they must be calibrated either directly or indirectly against one of the standard instruments of the IPTS.

Table 1   The International Practical Temperature Scale (IPTS)

Fixed point	Assigned temperature		Standard instrument
			optical pyrometer (above 1337.58 K)
Freezing point of gold	1337.58 K	1064.43 °C	
Freezing point of silver	1235.08 K	961.93 °C	
Freezing point of zinc	692.73 K	419.58 °C	thermocouple
Boiling point of water	373.15 K	100 °C	(903.89 K–1337.58 K)
Triple point of water	273.16 K	0.01 °C	
Boiling point of oxygen	90.188 K	−182.962 °C	
Triple point of oxygen	54.361 K	−218.789 °C	platinum resistance
Boiling point of neon	27.102 K	−246.048 °C	thermometer
Boiling point of equilibrium hydrogen	20.28 K	−252.87 °C	(13.81 K–903.89 K)
Equilibrium between the liquid and vapour phases of equilibrium hydrogen at 33 330.6 Pa pressure	17.042 K	−256.108 °C	
Triple point of equilibrium hydrogen	13.81 K	−259.34 °C	

Temperature measurement using changes in volume, length or pressure

2.1 Bimetallic strip

(a)

(b)

A bimetallic strip is one of the simplest examples of how the thermal expansion of two materials can be used to measure temperature. Strips of two different metals are bonded together as in Figure 3(a). The metals are chosen so that if they were allowed to expand freely, they would expand by different amounts for a given temperature change. Since they are constrained to be the same length at their common surface, a temperature change forces the combined strip to bend into a circular arc. If one end of the strip is fixed, the temperature change can be calibrated in terms of the deflection of the other end. Bimetallic strips made in the shape of a spiral are used directly as thermometers, with a pointer attached to the free end of the spiral (Figure 4).

Figure 3   A bimetallic strip at two different temperatures

The fractional increase in length per unit rise in temperature of a metal (when allowed to expand freely) is called its *coefficient of linear expansion*. Naturally, the greater the difference between the coefficients of linear expansion of the two metals in a bimetallic strip, the more the strip will bend for a given rise in temperature. For this reason, one of the metals is usually invar (an alloy of iron and nickel), which has a very low coefficient of linear expansion.

coefficient of linear expansion

> In Figure 3, if (a) represents the low-temperature situation and (b) the higher temperature, is the metal shown shaded pink the invar or the other metal?
>
> The invar is the metal shaded pink.

2.2 Liquid-in-glass thermometer

A liquid-in-glass thermometer also indicates temperature because of a *differential expansion*. In a mercury-in-glass thermometer the volume of the mercury increases with increasing temperature. The volume enclosed by the glass also increases, but by a smaller amount, because the coefficient of volume expansion is smaller for glass than for mercury, so the apparent length of the mercury column must increase. Thus the calibration of a liquid-in-glass thermometer depends upon the thermal expansion coefficient of the glass as well as that of the liquid it encloses.

differential expansion

If such a thermometer is taken from a cupboard at room temperature and plunged into boiling water, obviously the length of the mercury column does not increase instantaneously to its final value.

> Would you expect the rate of increase to be greater at first, when the temperature of the mercury is much below the temperature of the water, or later, when the mercury has nearly attained the temperature of the water?

It increases most rapidly when the thermometer is first placed in the hot water, because the rate of increase in length is proportional to the difference between the temperatures of the water and the thermometer. This subject is discussed further in Section 6 and also in the radio programme 'First-order response'. For the moment, notice that there is a delay from the time the thermometer is placed in a fluid until it indicates the temperature of that fluid to within the minimum error of which it is capable. This delay must be taken into consideration when deciding whether or not a thermometer is suitable for a given application. (In fact, the response time of *any* transducer is one of the major properties to be considered when choosing transducers.)

2.3 Sealed-fluid thermometers

Two other types of temperature-sensing device using a sealed fluid are shown in Figure 5. These can display a temperature reading at a distance from the point at which it is measured. In Figure 5(a) a liquid, such as mercury, completely fills the bulb, capillary and pressure sensor, which are made of metal. The temperature is sensed at the bulb. The capillary, which may be as much as 50 m long, connects the bulb to a pressure sensor such as a Bourdon tube. An increase in the temperature of the bulb causes the fluid in it to expand. This increased volume corresponds to an increased pressure on the Bourdon tube, which thus uncoils as explained in Unit 1. The resulting motion of the free end is used to drive a pointer.

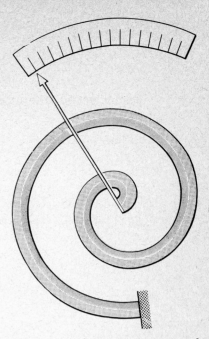

Figure 4   *A schematic representation of a thermometer using a coiled bimetallic strip*

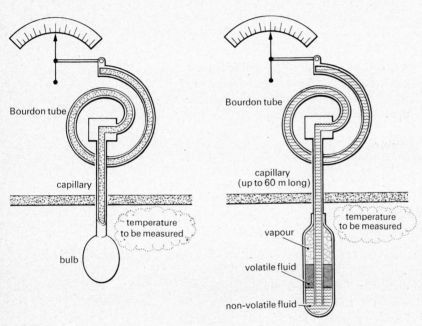

Figure 5   *Two types of thermometer using sealed fluid: (a) uses liquid expansion; (b) uses vapour pressure to sense temperature*

The temperature of the capillary tube and the Bourdon tube will also affect the transducer reading. Can you see how?

Any change in the temperature of the capillary or of the Bourdon tube from that at which it was calibrated will cause the volume of fluid enclosed to change, leading to an error in the reading. This effect can be minimized by making sure that the volume of the bulb is considerably greater than that of the remainder of the system.

The device in Figure 5(b) is constructed so as to avoid this problem. The sensing bulb contains a volatile liquid (i.e. one which vaporizes easily) and its vapour. The pressure of the vapour depends only upon the temperature in the bulb. The pressure is transmitted to the Bourdon tube by a non-volatile liquid. Changes in the temperature of the capillary tube or the Bourdon tube result in a change in the volume of the vapour in the bulb, but not in its pressure. Thus they do not affect the reading.

13

Thermoelectric (thermocouple) temperature sensors

3.1 Basic principles

A thermocouple is a temperature transducer consisting of two wires of different metals joined at both ends. If the two junctions between the metals are at different temperatures an electric current will flow around the circuit (see Figure 6). This phenomenon is called the *Seebeck effect* or sometimes the *thermoelectric effect*.

Seebeck effect
thermoelectric effect

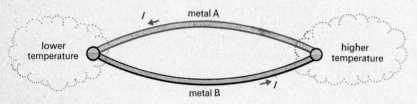

Figure 6   The Seebeck effect

If a voltmeter is inserted into the circuit as in Figure 7, it will indicate a voltage which depends upon the difference in temperature between the two junctions. If the voltmeter's internal resistance is sufficiently high, the voltage it displays is primarily due to the following effect.

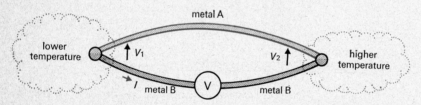

Figure 7   A basic thermocouple circuit

Across the junction of any two dissimilar metals there always appears a difference in electric potential called the *contact potential*. In Figure 7 the contact potentials are labelled V_1 and V_2. The contact potential between two metals varies with the temperature at their junction, increasing in magnitude with increasing temperature. The voltmeter reading equals the difference between the two contact potentials $V_2 - V_1$.

contact potential

Suppose that the temperatures of the two junctions were equal. What would the voltmeter reading be?

Zero. The two contact potentials would be the same.

This last conclusion, although quite obvious, has some useful implications. For although Figure 7 shows only two junctions between dissimilar junctions, unless the voltmeter and the wires leading to it are also made of metal B, there will be at least two more junctions of dissimilar metals. Figure 8 shows the two additional contact potentials V_3 and V_4. However, if the two connections between metal B and the copper wire are at the same temperature, V_3 and V_4 will be equal. The net voltage across the voltmeter will still be $V_2 - V_1$ because V_3 and V_4 are in opposite directions in the circuit.

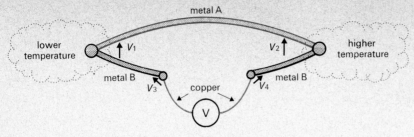

Figure 8   A thermocouple circuit showing the contact potentials at the meter connections

Figure 9 shows an alternative way of connecting a voltmeter into a thermo-couple circuit. The same reading will appear across the voltmeter in this arrangement as in Figure 8 if the temperatures at the junctions are the same as in Figure 8. Now the voltmeter reading will equal the difference between the net contact potential $V_5 - V_6$ at the low-temperature junction and the contact potential V_2 at the high-temperature junction. The net low-temperature contact potential $V_5 - V_6$ will be the same as the contact potential which would appear if metals A and B were connected together directly (as in Figure 8) because the third metal, copper, is common to both of the low-temperature connections.

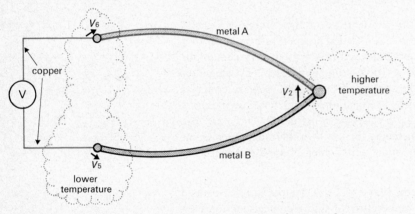

Figure 9   An alternative way of connecting the meter in a thermocouple circuit

The results of the preceding discussion can be summarized in the following rule of thermocouple operation:

Rule 1.   If a third metal is introduced into a thermocouple circuit, the net voltage is not affected, provided both junctions of the third metal are at the same temperature.

Rule 1 explains why a voltmeter can be inserted into the thermocouple circuit in either of the ways shown in Figures 8 and 9. It also tells us that thermocouple wires can be brazed or soldered, and that any new metals introduced in the process will not alter the net thermocouple voltage.

SAQ 1

A voltmeter need not necessarily be connected to the colder of the two junctions in a thermocouple circuit. Suppose it is connected to the hotter junction, as in Figure 10. If the temperatures of the junctions are the same in Figures 9 and 10, how will the contact potentials V_A, V_B and V_C and the voltmeter reading compare with the contact potentials and voltmeter reading in Figure 9?

3.2   Reference junctions

When a thermocouple is used to measure temperature, one of its two junctions is placed in contact with the material whose temperature is to be

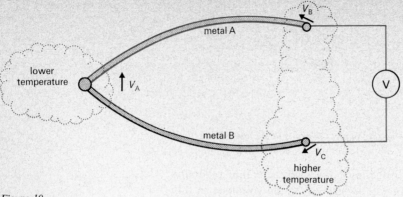

Figure 10

measured so that it takes on the temperature of that material. However, as I have said in the previous section, the voltage across the output terminals of a thermocouple depends upon the temperature at *both* junctions. This means that the temperature at the second junction must be either known or controlled to at least the accuracy expected from the measurement. The second junction is usually referred to as the *reference junction*.

One strategy for establishing the temperature of the reference junction is to place it in some apparatus in which the temperature is accurately controlled. The most accurate method, used primarily in very demanding laboratory situations, is to use a 'triple-point apparatus', in which the temperature is maintained at the triple point of water. A more common method is to use an ice–water mixture to maintain a reference temperature of 0 °C. Such an ice bath can quite easily be constructed. There are also ice baths which are made commercially for use with thermocouples. Some of these contain built-in refrigeration units which eliminate the need to replenish the ice.

The reference temperature need not be 0 °C. A temperature-controlled oven can be used to keep the temperature of the reference junction constant at any convenient temperature.

With the temperature at the reference junction established by some suitable means, the temperature at the measuring junction can be determined by measuring the voltage across the thermocouple terminals. The meter used to measure this voltage may have a scale which is calibrated directly in degrees Celsius (or in kelvins or degrees Fahrenheit). Alternatively, the measurement can be made in volts and converted to temperature by using standard thermocouple tables. The table for a thermocouple (of a given type of wires) gives the voltage output of the thermocouple at different temperatures of its measuring junction, for a given reference junction temperature. The latter is usually 0 °C.

If you are using a thermocouple with a reference junction at a temperature other than 0 °C, yet wish to use the standard table, this can be done very easily. You simply look up the voltage reading for the reference temperature you are using in the table and add it to the actual reading you get. The sum is the voltage you would have measured using a reference temperature of 0 °C. This procedure is based on a second rule of thermocouple operation.

As you can see from Figure 11, this rule follows directly from the fact that the thermocouple voltage is the difference between two contact potentials, both of which appear across a junction of the same two metals.

Rule 2. (See Figure 11.) *Suppose a thermocouple has an output voltage of V_{AB} when its junctions are at temperatures T_A and T_B, and an output*

16

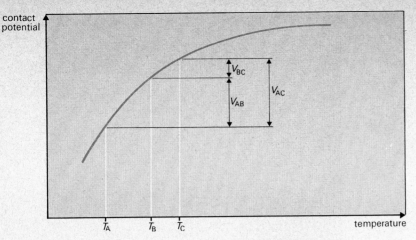

Figure 11   An illustration of Rule 2 of thermocouple operation

voltage of V_{BC} when its junctions are at T_B and T_C. Then its output voltage will be $V_{AB} + V_{BC}$ when its junctions are at T_A and T_C.

SAQ 2

A thermocouple with the table shown below has a voltage of 5.13 mV when its measuring junction is placed in boiling water. At what temperature is the reference junction?

If the voltage is 3.15 mV, what temperature is being measured?

Chromel–constantan thermocouple (reference junction at 0 °C)

Temperature/°C	Potential/mV
20	1.19
40	2.41
60	3.68
80	4.99
100	6.32
120	7.69

Extension wires

In many instrumentation situations, the measuring instruments must be located far from the point at which the measurements are to be made. With thermocouples this raises the problem of where to put the reference junction (which is located where the thermocouple wires join the wires leading to the meter). If it cannot be located near where the measurement is to be made, long and expensive lengths of thermocouple need to be used. The usual alternative is to use extension wires made either of the same material as the thermocouple wires, but of a smaller gauge and to less exacting standards, or made of some other metal which closely matches the temperature characteristics of the thermocouple wires. With extension wires connected, the reference junction is effectively moved to their ends.

In what ways must the extension wires match the thermocouple wires?

There must be no contact potential at the junction of each extension wire and the corresponding thermocouple wire. Also, the contact potential between the two extension wires at the reference junction must be the same as the contact potential that would have been generated by the thermocouple wires at the reference junction.

The extension wires need only match the thermocouple wires over the range of temperatures that the extension wires are exposed to, that is, the temperature at the reference junction and at the junctions of the thermocouple

17

SPECIFICATION

**COMARK TYPE 1601
ELECTRONIC THERMOMETER**

Meter
120 mm scale.

Input
Terminals on front panel: 4 mm socket.

Thermocouple materials
Nickel–chromium and nickel–aluminium to British Standard 1827.

	Range	Resolution/division
from −87 °C to +25 °C		2 °C
from 0 °C to 100 °C		1 °C
from 0 °C to 300 °C		5 °C
from 0 °C to 1000 °C		10 °C

Accuracy
(at 23 °C) ±2% f.s.d.

Automatic cold junction
Ambient change from 0 °C to 40 °C gives less than ±2 °C deviation.

Input lead resistance
1000 Ω causes 1 °C error.

D.C. output
+1 V for f.s.d. 2 mA maximum current.

Choice of power supplies
Internal battery or 110 V or 240 V, 40–60 Hz.

Figure 12   Comark type 1601 electronic thermometer using a thermocouple

and extension wires. The temperature range to which the measuring junction of the thermocouple is exposed may be much greater.

3.3 Transducer example 1: A thermocouple electronic thermometer

Figure 12 shows an electronic thermometer which uses a thermocouple probe. In this instrument, the temperature of the reference junction is not held constant. Instead, it is allowed to take on the ambient temperature. The ambient temperature is measured using a thermistor (which will be described in section 4.4), and, based on this measurement, an appropriate voltage is generated and electronically added to the thermocouple voltage. That is, the voltage which is added depends upon the ambient temperature. The total voltage is amplified and fed to the meter, which is calibrated in degrees Celsius. In this way the instrument has what is called an 'automatic cold junction'.

This method of cold-junction compensation is a direct application of the principle of Rule 2. This principle is also used to obtain the range of temperatures below 0 °C which you will see listed in the specification. To obtain this range, a second voltage is added to the thermocouple voltage, in addition to the reference-junction compensating voltage. With this extra voltage included, the output voltage from the thermocouple is the same as it would be if the temperature of the reference junction were −87 °C.

3.4 Thermocouple materials and construction

Since thermocouples have been widely used for temperature measurements for many years, various standards for thermocouples have been developed and adopted by international and national standards bureaux. The metals or alloys of metals used for the most common thermocouples are listed in Table 2. For each of the types listed there, there are published tables giving their output voltages for a wide range of temperatures at their measuring junctions, and giving allowable tolerances. Thus now, most manufacturers of the listed types of thermocouple manufacture them to meet the requirements of the standard. This not only ensures that the table accurately describes the behaviour of the thermocouple but permits interchangeability between thermocouples of different makers.

The characteristics of temperature against voltage for these six thermocouple materials are shown graphically in Figure 13. Although at first

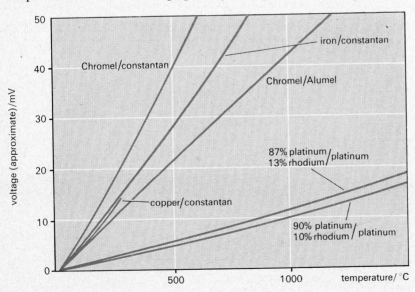

Figure 13   Curves of voltage against temperature for several common thermocouple materials

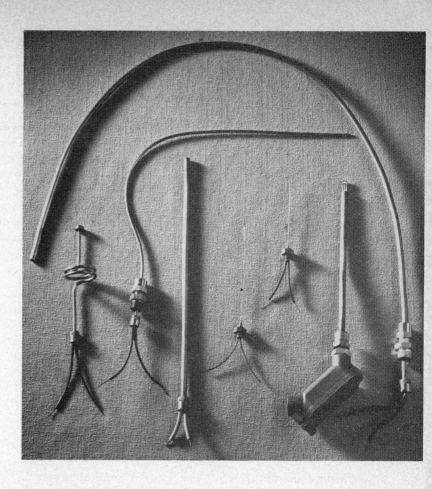

SPECIFICATION

PYROTENAX MINERAL-INSULATED THERMOCOUPLE PROBES

Response time of mineral-insulated thermocouple probes

Nominal probe diameter/mm	Time constant/s
1.0	0.15
1.5	0.24
2.0	0.35
3.0	0.90
4.5	1.35
6.0	3.90

Combinations of probe wire and sheath material and their uses

Thermocouple material	Sheath material	Maximum temperature/°C	Typical applications
copper–constantan	cupro-nickel	400	boiler flue gas, food processing, subzero temperatures, plastic moulding, etc.
iron–constantan	mild steel	600	paper and pulp mills, polyethylene manufacture, tar stills, etc.
iron–constantan	chromium–nickel–titanium stabilized steel	750	chemical reactors, perfume distilling, re-heat and annealing furnaces, etc.
nickel chromium–nickel aluminium	chromium–nickel–titanium stabilized steel	800	boiler steam and water, footwear mouldings, nuclear energy 'in-pile' instrumentation, etc.
nickel chromium–nickel aluminium	chromium–nickel steel	1100	blast-furnace gases, brick kilns, glass manufacture, etc.

Figure 14   Pyrotenax mineral-insulated thermocouple probes

Table 2 Thermocouple materials

Designated type and British Standard (BS) number	Metal or alloy for 1st wire	Metal or alloy for 2nd wire
Type S* (BS 1826)	platinum	90 per cent platinum, 10 per cent rhodium
Type R (BS 1826)	platinum	87 per cent platinum, 13 per cent rhodium
Type J (BS 1829)	iron	constantan (57 per cent copper, 43 per cent nickel plus small amounts of other materials)
Type T (BS 1828)	copper	constantan
Type E	Chromel (90 per cent nickel, 10 per cent chromium)	constantan
Type K† (BS 1827)	Chromel	Alumel (94 per cent nickel, 3 per cent manganese, 2 per cent aluminium, 1 per cent silicon)

*The Type S, or Le Chatelier, thermocouple is used to define the International Practical Temperature scale between 903.89 K and 1337.58 K.

† Both Chromel and Alumel are brand names, belonging to one manufacturer. Other manufacturers who make thermocouples to BS 1827 often refer to them as nickel chromium–nickel aluminium (as in Transducer examples 1 and 2) or as Type K thermocouples.

glance the curves appear linear, if you check with a ruler, you will find that they are not. The curves also show the great differences in sensitivity between the different thermocouple types, with the Chromel–constantan thermocouple producing about ten times the output of the platinum–platinum 10 per cent rhodium thermocouple. The thermocouple types also differ greatly in their abilities: (a) to withstand corrosive environments of different types and over different temperature ranges; (b) to withstand thermal and mechanical stresses; (c) in their accuracy and reproducibility; and of course they also differ in price. The particular combination of attributes necessary determines the choice in any given application.

3.5 Transducer example 2: A range of thermocouples

A thermocouple probe at its simplest consists only of the two wires with their ends soldered or welded to form a junction. In many applications that is all that is needed. The junction of the bare wires is placed in contact with whatever is to have its temperature measured and the measurements are taken.

The thermocouple probes described in this section are more elaborate than that. Figure 14 shows a selection of them. The actual thermocouple wires and the measuring junction are encased in a metal sheath, as you can see in the X-ray photographs in Figure 15. Between the wires and the sheath is a layer of magnesium oxide insulation. For this reason, this type of probe is called a *mineral-insulated thermocouple probe*. In the probe on the left in Figure 15 the measuring junction touches the metal sheath, while the probe on the right is insulated from it.

Figure 15 X-ray photographs of mineral-insulated thermocouple probes

mineral-insulated thermocouple probe

The function of the metal sheath is to protect the thermocouple from environments in which it might be corroded, abraded or otherwise damaged. Together with the seal at the end, the sheath also enables the thermocouple to be used in a system which must be sealed or must withstand high pressures.

The magnesium oxide insulation provides an insulation resistance of not less than 1000 MΩ between the thermocouple wires and the metal sheath. At the same time it has a high thermal conductivity to enable the probe to respond rapidly to changes in the temperature to be measured.

The response rate of the thermocouple depends upon the rate of heat transfer from the material being measured and so varies from one application to another. However, to give you some idea, the manufacturer has supplied the information in the specification. The probe was plunged into stirred water at 90 °C. Figure 16 shows an oscilloscope trace showing a typical probe output voltage resulting from this experiment. Traces such as these were used to determine the 'time constant' listed in the specification, which is the time taken for the temperature reading to rise to 63 per cent of its final value. (The meaning of time constant is explained more fully in Section 6.)

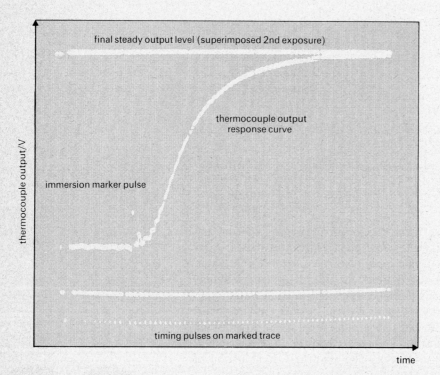

Figure 16   An oscilloscope trace showing the response of a thermocouple probe to a rapid immersion in water at 90 °C

The materials used for both the sheath and the thermocouple wires depend upon the temperature range and application. The specification lists several of the proble and wire materials used by this manufacturer.

Section 4

Resistive temperature transducers

4.1 Basic principles

In Unit 2 you examined the way in which the resistance of a strain gauge changes when the gauge is strained. Some of the techniques described in that unit showed how to compensate for changes in the resistance of the strain gauge due to changes in its temperature. This section also discusses resistive transducers. Now, however, the changes in resistance due to temperature are just what we are looking for. Moreover, we have to ensure that there are no changes in resistance due to strain to confuse our measurement. Some of the measuring techniques in this section will already be familiar to you from Unit 2.

As with strain gauges, resistive temperature transducers can be divided into two major types: metals and semiconductors. The most commonly used semiconductor temperature transducers are made of oxides of chromium, manganese, iron, cobalt or nickel and are called *thermistors*. Their properties are very different from metallic resistance transducers, so I shall discuss them separately.

thermistor

Figure 17 shows how the resistance of platinum, copper, tungsten, Balco (an alloy of nickel) and nickel vary with temperature. The resistance shown is relative to the resistance R_0 at 0 °C. These are the metals most often used in resistive temperature transducers. A curve for a thermistor is also included for comparison.

> Which metal has the lowest sensitivity to temperature change and which has the highest sensitivity?
>
> ---
>
> Platinum has the lowest sensitivity; nickel has the highest.

Although the sensitivity of platinum is the smallest, it is the most stable and reproducible type of temperature transducer. It is thus used to define the International Practical Temperature Scale for temperatures in the range from 13.81 K to 903.89 K. Platinum is chemically inert and is resistant to contamination. This enables platinum resistance thermometers to be used in a wide range of environments. Unfortunately, platinum is relatively expensive.

Of the base metals used for resistance thermometers, nickel is the most widely used, primarily in the temperature range from about 175 K to 600 K.

To help quantify the sensitivity of these metals, the curves in Figure 17 can be approximated in the temperature interval between 0 °C and 100 °C by the linear relation.

$$R = R_0(1 + aT),$$

where a is called the *temperature coefficient of resistance*.

temperature coefficient of resistance

In Table 3 I have listed the values of a for some metals for the range from 0 °C to 100 °C.

Table 3   Temperature coefficient of resistance for the range from 0 °C to 100 °C

Material	$a/(\mathrm{K}^{-1})$
copper	0.0043
nickel	0.0068
platinum	0.0039
tungsten	0.0046

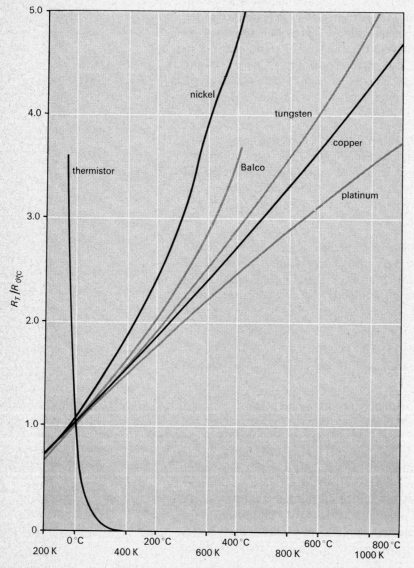

Figure 17   Relative resistance against temperature for various materials used in resistance thermometers (Sources of data. Thermistor: from General Electric Company technical literature; Nickel: from Sigmund Cohn technical literaure; Balco: REC experimental data; Tungsten: Metals Handbook, 8th edn, vol. 1, p. 1225, American Society for Metals, 1961; Copper: Forsythe, W. E., Smithsonian Physical Tables, 9th edn, Washington, DC; Platinum: mean of NBS calibrations of platinum thermometers)

SAQ 3

If the resistance of a length of platinum wire is 100 Ω at 0 °C, what is its resistance at 50 °C?

4.2 Construction of metallic resistance thermometers

Metallic resistance thermometers are made in a great variety of sizes, shapes and forms, depending upon the application for which they are designed. One common approach uses a sensing element mounted within a metal probe to protect it. Figures 18 and 19 show three ways in which the sensing element can be constructed. Figure 18 shows an 'open-wire' element. The coil of wire (platinum in this case) is wrapped around a supporting cage. It is exposed directly to the fluid whose temperature is to be measured, through the holes in the guard tube.

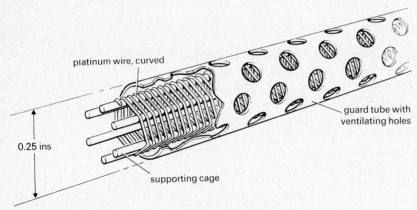

Figure 18   The construction of an 'open-wire' resistance sensor

Figure 19 shows two methods of construction in which the resistance wire is isolated from the fluid. This is to avoid the resistance wire becoming contaminated with other materials, since contamination would change the relationship of its resistance to temperature. The ceramic on which the resistance wire is mounted also helps the sensor to withstand mechanical shock. In all cases the construction is designed so that the least amount of strain is placed upon the resistance wire as a result of differences in expansion of the wire and its support.

Why is this necessary?

Strain in the resistance wire would cause a change in its resistance not due to temperature.

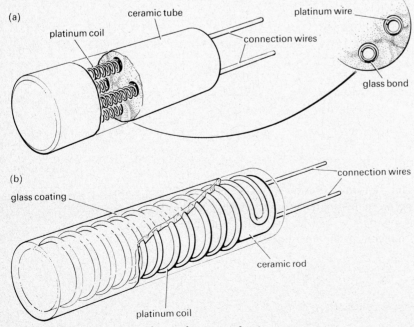

Figure 19   The construction of two other types of resistance sensor

25

Figure 20 shows a selection of probes in which resistive sensing elements are mounted. Apart from variations in size, they vary in the absence or presence of holes to enable the material whose temperature is being measured to contact the sensing element within.

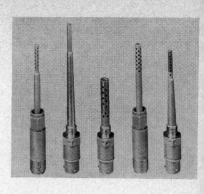

Figure 20 Probes for use with platinum resistance sensors

4.3 Transducer example 3: A platinum resistance temperature sensor

The platinum resistance temperature sensor in this example is shown, full size, in Figure 21. It is designed to be used either unsheathed or built into a thermometer element. Its internal construction is as shown in Figure 19(a).

Just as with thermocouples, there are established standards to which many platinum resistance thermometers are constructed. The relevant standard for these sensors is BS 1904:1964. It specifies the calibration of resistance values and tolerances against temperature which the sensor must meet. The sensors are classified according to their resistance at 0 °C and the change in resistance between 0 °C and 100 °C. This temperature interval is called the *fundamental interval* (f.i.). In this example, the sensors have a resistance of 100 Ω at 0 °C and a fundamental interval of 38.5 Ω. An extract from the resistance table for this calibration is shown in Table 4 to give you an idea of how the resistance varies with temperature.

fundamental interval

Table 4

Temperature/ °C	Maximum error ($\pm$)/°C	38.5 f.i. calibration	
		Resistance/Ω	Maximum error ($\pm$)/Ω
−200	1.2	18.56	0.5
−150		39.73	
−100	0.5	60.27	0.2
−50		80.31	
0	0.3	100.00	0.1
50		119.40	
100	0.5	138.50	0.2
150		157.32	
200	1.0	175.85	0.35
250		194.09	
300	1.4	212.05	0.5
350		229.72	
400	1.9	247.11	0.65
450		264.20	
500	2.4	281.01	0.8
550		297.53	
600	3.0	313.77	0.95

Some items are listed in the specification for Transducer 3.3. If an electric current flows through a resistor, it dissipates energy, which heats that resistor. This is what is meant by the 'self-heating effect' in the specification. The specification also mentions the 'thermoelectric effect'.

Why should this occur?

The platinum resistance wire is connected to copper lead wires, thus forming a thermocouple.

SPECIFICATION

ROSEMOUNT MODEL E712
PLATINUM RESISTANCE TEMPERATURE DETECTOR

Temperature range
The temperature detector is designed for use over the temperature range from − 200 °C to +800 °C.

Thermal response time
When quickly plunged into water at approximately +80 °C moving at 1 m s⁻¹ with flow transverse to the detector, 63 per cent of the response to the step of temperature shall take place in less than 0.4 s.

Insulation resistance
The detector, when mounted in a close-fitting metal sheath under dry conditions, shall have an insulation resistance between the leads and the metal sheath of greater than 10 MΩ at 240 V at ambient temperature.

Self-heating
When tested in accordance with BS 1904 :1964 section 3.16, the indicated temperature rise in the temperature detector with a power of 10 mW dissipated in it shall not exceed +0.3 °C.

Thermoelectric effect
Any thermoelectric potentials present in the detector shall not cause the measurement to shift outside the tolerance band specified by BS 1904 :1968.

Stability
When the detector is subjected to ten consecutive thermal shocks from ambient temperature to − 200 °C, and from +800 °C to ambient temperature, the resistance at 0 °C shall not change by more than ±0.05 per cent.

Figure 21   Rosemount model E712 platinum resistance temperature detector

4.4 Thermistors

In Figure 18 I included a curve for a thermistor along with the curves showing the resistance change of several metals with temperature. The thermistor curve is strikingly different from the other curves in two main respects: it is much steeper and it decreases rather than increases with increasing temperature. That is, its temperature coefficient of resistance is negative* and, at least over part of its range, is much larger in magnitude than that of any of the metals. The relationship between resistance and temperature for a thermistor can be described by the formula

$$R = A \exp \frac{B}{T},$$

where A and B are constants and T is the temperature on the thermo-dynamic scale. B is called the *characteristic temperature* of the thermistor. Its value is usually between 2000 K and 4000 K. You will see how useful it is in the transducer example of section 4.6.

characteristic temperature

To give you an idea of what this formula means numerically, I have selected a set of values for one particular thermistor in Table 5. Notice that over the range from 0 °C to 300 °C the resistance has decreased by a factor of about 3000! Contrast this with a similar set of values in Table 4, where the resistance of a platinum resistance thermometer increased by a factor of about 2. The thermistor resistance changes very greatly, but it also changes very non-linearly. The resistance change in the 50 °C interval above 0 °C is 320 000 Ω, while in a similar interval above 250 °C it is 130 Ω.

Table 5 Selected resistance values for a thermistor

$T/°C$	R/Ω
0	355 000
25	100 000
50	34 000
100	6000
150	1650
200	580
250	240
300	110

There is an alternative way of expressing the thermistor formula which is useful if the value of the resistance is known at some temperature T_0. If the resistance at that temperature is called R_0, then

$$R_0 = A \exp \frac{B}{T_0}.$$

Dividing the previous equation by this one and simplifying gives

$$R = R_0 \exp B \left[\frac{1}{T} - \frac{1}{T_0} \right].$$

---

* There are also thermistors with positive temperature coefficients of resistance. However, they are used primarily in applications such as overheating protection, rather than for temperature measurement.

28

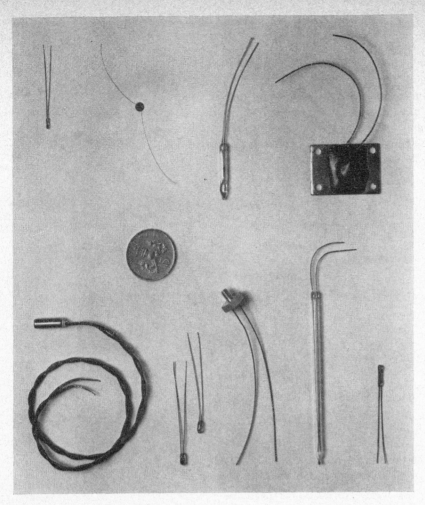

Figure 22   Several types of thermistor

Figure 22 is a photograph of several types of thermistor used for temperature measurement. As temperature-sensing elements they can be made
much smaller than metallic resistance sensors. This enables them to
respond to changes in temperature more quickly. However, this also means
that their self-heating effect is greater for the same current than metallic
sensors, and so they must be operated at lower current levels.

SAQ 4   SAQ 4

A thermistor has a characteristic temperature B of 3000 K. If its resistance is 100 kΩ at 300 K, what will its resistance be at 600 K?

(Note: $\exp(-5)=0.0067$.)

4.5   Measuring circuits for resistive temperature transducers

In Unit 2 you studied circuits for measuring changes in the resistance of a
strain gauge. Measuring the resistance change of temperature transducers
is similar, but there are some major differences.

What is the major difference in the size of the resistance change between
the two types of transducer?

The change in resistance of a strain gauge is always a small fraction of the
total resistance of the gauge. By contrast, in the last section I mentioned
a thermistor which changed its resistance by a factor of 3000 over a 300 K

29

interval and a platinum resistance thermometer which changed resistance by a factor of 2 over the same temperature interval.

This relatively large change of resistance in resistive temperature transducers has important implications for the design of bridge circuits used with them. However, it also enables other circuits, such as the one shown in Figure 23, to be used. The electrical source is what is called a *constant current source*. Ideally, a constant current source supplies the same current no matter what voltage appears across it. In practice, of course, the current does vary somewhat.

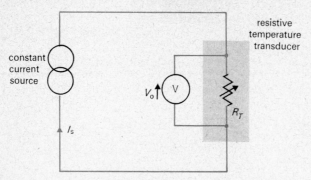

Figure 23   *A measuring circuit for resistance temperature sensors*

With a current I_s flowing through the resistive temperature transducer, the voltage across it will be

$$V_o = R_T I_s,$$

where R_T is the resistance of the transducer. Since I_s is held constant by the constant current source, V_o is directly proportional to the transducer's resistance.

Bridge circuits

The most common measuring circuit used with resistive temperature transducers is the bridge circuit. Consider the circuit shown in Figure 24. Because this is a circuit for measuring temperature, the wires leading from the transducer to the bridge circuit will have a temperature gradient across them. Moreover, this temperature gradient will change as the temperature being sensed changes.

> How will this affect the voltage reading across the output terminals of the bridge?

The bridge will measure not only the change in resistance of the temperature sensor, but also the change in resistance of the wires leading to it. Thus the bridge voltage will not accurately reflect changes in the resistance of the temperature sensor.

The two circuits in Figure 25 use extra wires leading to the transducer to help compensate for this effect. In the three-wire circuit in Figure 25(a)

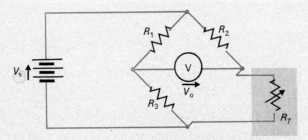

Figure 24   *A bridge circuit for resistive temperature sensors*

the outer two wires are in different arms of the bridge. Thus, if they both have the same resistance, they will not affect the bridge output voltage. The middle wire is in the meter circuit and not in any arm of the bridge. If the current through the meter is much smaller than the current through the bridge arms, the resistance of this lead will have a negligible effect on the voltmeter reading.

In the four-wire circuit in Figure 25(b), the two extra wires form a compensating loop. Their resistance varies in the same way as does the resistance of the transducer leads, but this change of resistance appears in the adjacent arm of the bridge.

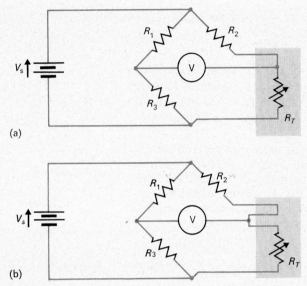

(a)

(b)

Figure 25 These bridges use three-wire and four-wire connections to compensate for resistance changes in the leads

SAQ 5

SAQ 5

In the circuit of Figure 23, how will changes in the resistance of the transducer leads affect the voltmeter reading?

(Assume that the voltmeter current is negligible.)

Now let us consider the effect of the large resistance change in temperature transducers. In Unit 2 you found that when a bridge circuit is used to measure resistance change in one arm by measuring the out-of-balance voltage, this voltage is not linearly related to the resistance. In a strain-gauge bridge this non-linearity is often small enough to be neglected because the resistance change is only a small fraction of total strain-gauge resistance.

There are various methods used to reduce the non-linearity of a bridge circuit. One straightforward method is to divide the desired temperature range into several smaller ranges. For example, in Figure 26, if $R_1 = R_2$, then choosing values of R_3 equal to the value of R_T at the midpoint of each temperature range desired will make the bridge balance at these temperatures. By switching in the appropriate values of R_3 for each range, the non-linearity will be much lower than if one value of R_3 is used for all three ranges.

From a similar circuit in section 5.1 of Unit 2, you may remember that, if the current drawn by any measuring circuit connected across the bridge output can be neglected, the bridge output voltage is

$$V_o = \left[\frac{R_T}{R_T + R_2} - \frac{R_3}{R_3 + R_1} \right] V_s.$$

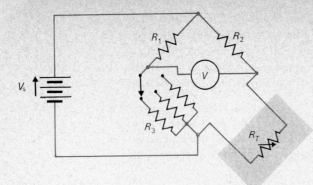

Figure 26 A bridge circuit with several ranges for measuring the resistance of a temperature sensor

If R_T did not appear in the denominator of the first term, this relation would be linear. Thus another strategy to linearize the bridge is to make R_2 and R_1 much larger than the midscale value of R_T (while still keeping them equal). This, however, reduces the sensitivity of the bridge.

> *A small reminder.* Lest you are misled by the preceding discussion, remember that a non-linear response does not necessarily mean an inaccurate response. While in some applications, especially those involving control, a linear response may be needed, in others a non-linear response with a correspondingly non-linear meter scale or an appropriate conversion table might be perfectly acceptable.

SAQ 6

Why does increasing R_1 and R_2 make the bridge more linear but reduce the sensitivity?

SAQ 6

4.6 Transducer example 4: An electronic thermometer using a thermistor

In contrast to the other transducer examples in Units 3 and 4, this example is not a complete, commercially available package. It is taken from a manufacturer's application note describing the construction of 'A 0 °C to 50 °C Electronic Thermometer', which incorporates a thermistor as sensing element. I have included it because it illustrates some interesting refinements in a simple bridge circuit.

Figure 28 is a photograph of the thermistor recommended for use with this thermometer. The actual resistance element is the black dot near the tip of the glass envelope. Figure 27 shows the bridge circuit used with this

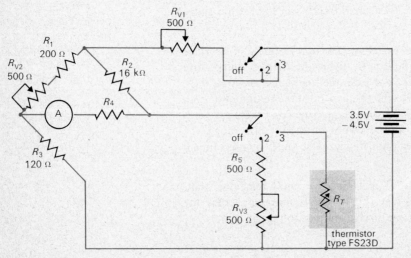

Figure 27 The measuring circuit for transducer example 4

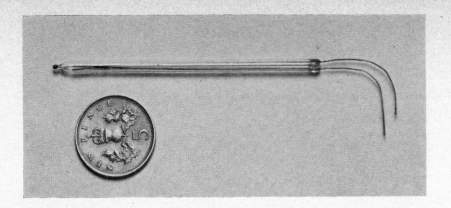

SPECIFICATION

ITT TYPE FS23D THERMISTOR

Resistance
At 20 °C (approx.) : 2 kΩ;
At 25 °C (nominal) : 1680 Ω ±20%;
At 300 °C (approx.) : 13 Ω.

Temperature coefficient of resistance
At 20 °C : −3.6%/°C (approx.)

Characteristic temperature B
25 °C—85 °C, nominal : 3050 K ±5%.

Length
76.2 ±3.2 mm.

Maximum ambient temperature
(Provided that the temperature of the leads does not exceed the melting point of the tinning which is 175 °C) : 300 °C.

Maximum bead temperature
(Provided that the maximum power dissipation is not exceeded) : 300 °C.

Maximum mean power dissipation in free air at 20 °C
Averaged over any 20 ms period : 360 mW.

Maximum instantaneous power dissipation in free air at 20 °C
720 mW.

Maximum recommended power dissipation for temperature measurement
10 mW.

Nominal thermal cooling time constant τ **in free air from the self-heated state**
20 s.

Mass
(Nominal) : 1.8 g.

Figure 28   ITT type FS23D thermistor

thermistor to measure temperature. To understand the details of this circuit we need to know the range of resistance values of the thermistor which correspond to a temperature range from 0 °C to 50 °C.

What characteristics of the thermistor do we need in order to calculate its resistance values?

We need its characteristic temperature B and its resistance R_0, at one temperature T_0. Then we can use the formula

$$R = R_0 \exp B \left[\frac{1}{T} - \frac{1}{T_0} \right]$$

to calculate R at 0 °C and 50 °C. The manufacturer specifies B as 3050 K ± 5 per cent and gives a resistance value of 1680 $\Omega \pm 20$ per cent at 25 °C. Using these numbers, I calculated the resistance at 0 °C and 50 °C as

R at   0 °C $= 4300$ Ω,
R at 50 °C $=$ 760 Ω.

You can see from the tolerances on B and the resistance at 25 °C that the two calculated values of resistance are only approximate. The resistance values of the bridge must be adjusted to suit the particular thermistor used. That is why the variable resistors, R_{v1} and R_{v2} are needed. With the switch in position 3 and the thermistor held at 0 °C, R_{v2} is adjusted until the bridge is balanced. Then, with the thermistor held at 50 °C, R_{v1} is adjusted until the meter gives a full-scale reading. R_{v1} can also be adjusted to compensate for changes in the battery voltage due to ageing. To do this conveniently, without having to subject the thermistor to an accurate 50 °C temperature each time R_{v1} is to be adjusted, R_5 and R_{v3} are set so that their resistance is equal to the thermistor resistance at 50 °C. This is done once, when the thermometer is first calibrated. Thereafter, to adjust R_{v1} for changes in battery voltage, it is simply necessary to set the switch to position 2 and adjust R_{v1} for full-scale deflection.

What are some of the sources of non-linearity in this thermometer?

The resistance of the thermistor varies exponentially, not linearly, and since the thermistor resistance changes by a factor of five over the range from 0 °C to 50 °C, the bridge output varies non-linearly with thermistor resistance. Fortunately, however, these two sources of non-linearity can be made to cancel each other partially. The component values, including the loading effect of the meter, were chosen to maximize this cancellation. The meter has a full-scale deflection of 250 µA and must have a resistance of less than 1200 Ω. Resistor R_4 is chosen so that the combined resistance of it and the meter is about 1200 Ω, the value which gives the best linearity for this circuit. The resulting deviation from linearity is 0.5 °C or less over the entire temperature range, as shown in Figure 29.

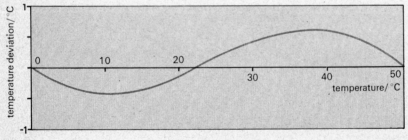

Figure 29   *The deviation of the bridge output from linearity (assuming zero circuit tolerances)*

Section 5

Radiation pyrometers

All the temperature sensors described so far rely upon physical contact between the sensor and the object being measured to ensure that the sensor and object are at the same temperature. In this section we shall look at a class of devices called *radiation pyrometers* which measure temperature without any contact between measured object and sensor. Figure 30 shows one such device. The operator looks through a lens in the sensing head to aim it at the object whose temperature is being measured, which is referred to as the target. The size of the target and its distance from the sensing head are not important so long as the image the operator sees is greater than a certain minimum size, as I shall explain in section 5.4. The temperature of the target is displayed on a meter on the electronic unit which covers temperatures from −50 °C to 650 °C in several ranges. Basically, the instrument works by focusing infrared radiation emitted by the target onto a small resistance thermometer inside the sensing head. This radiation heats up the resistance thermometer. An electrical circuit converts the rise in temperature of the resistance thermometer into an electrical signal which is calibrated in terms of the temperature of the target.

radiation pyrometers

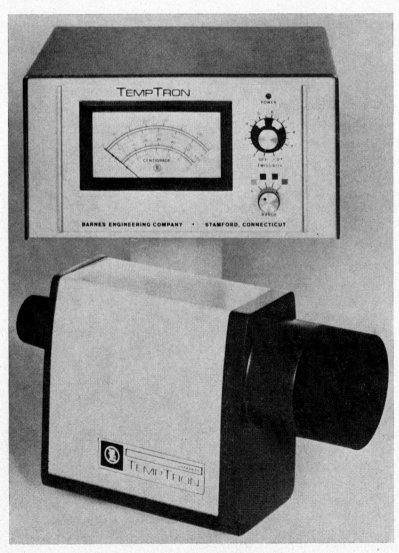

Figure 30 TempTron infrared radiation thermometer

35

In the remainder of this section I shall cover the operating principles of radiation pyrometers in more detail. As you can see from the example, there are a number of problems to be investigated:

1 What is the nature of the radiation emitted by the target and how does it depend upon temperature?

2 Why is the measurement independent of the size of the target and its distance from the sensing head?

3 What techniques, mechanical and electrical, are used to convert the focused radiation into a signal accurately calibrated in terms of the target temperature?

5.1 Infrared radiation

In everyday life we encounter various forms of radiated energy: visible light, infrared radiation from fires, and radio waves, for example. These three apparently different types of energy are in fact all forms of *electromagnetic radiation*. Such radiation can be characterized by its *wavelength* and its *frequency*.

> If you picture wave motion in terms of water waves, the wavelength is the distance between successive peaks and the frequency is the number of peaks passing a given point every second. Thus multiplying the wavelength by the frequency of a wave gives the speed at which it propagates.

The only difference between radio waves, infrared radiation and light is in their frequency f and their wavelength λ. They all propagate at the speed of light, which in free space is $c = 300$ Mm s^{-1}. Figure 31 shows the names we normally use to describe electromagnetic radiation at different frequencies and wavelengths. The only part of the electromagnetic spectrum to which the human eye responds is in the wavelength range 0.4–0.75 μm, which is marked as the visible band in Figure 31.

Within the visible band we can differentiate between radiation of different wavelengths. Differences in wavelength appear to us as differences in colour. That part of the electromagnetic spectrum just below the low frequency (or red) end of the visible band is called the infrared band. Although we cannot see infrared radiation, it can still be focused and otherwise handled by an optical system in just the same way as visible light. The infrared band is the part of the electromagnetic spectrum most used in radiation pyrometers.

> The discovery of infrared radiation is quite an interesting story. In 1800, using a prism to obtain a spectrum of visible light, William Herschel explored the heating ability of the different colours of the spectrum by using a thermometer. He moved the thermometer along from colour to colour in the spectrum recording the temperature the thermometer reached. But to his surprise, the most effective place to heat the thermometer was actually *beyond* the red end of the visible spectrum.

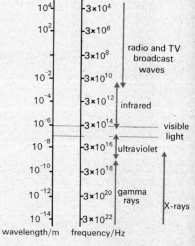

Figure 31   The electromagnetic spectrum

5.2 Black bodies and their properties

The radiation pyrometer in Figure 30 can be used to measure temperatures as low as −50 °C. Thus it is clear that infrared radiation is emitted not only by 'hot' objects like electric fires or the sun. In fact, any object whose temperature is not absolute zero (0 K) emits radiation. The amount it emits depends upon its temperature. At the same time it receives radiation from everything around it. Some of this radiation is absorbed, but some is reflected and some may be transmitted through the object. (For example, glass and air transmit nearly all the radiation in the visible band which is incident upon them.)

As we are concerned with the radiation which is emitted, let us start by considering objects which do not reflect or transmit any of the energy incident upon them. Such objects must absorb *all* incident energy. They are called *black bodies*. Because they also emit radiation, black bodies do not necessarily appear black; the term is merely a technical one.) Although there are no perfect black bodies, it is important to understand their properties, as several of the techniques used with radiation pyrometers are designed to make the target appear to the pyrometer to be a black body.

black bodies

When you use a prism to split the sun's radiation into the colours of the visible spectrum, you are demonstrating that the radiation contains a range of wavelengths.

The significance of Herschel's discovery of infrared radiation is that the sun also radiates at wavelengths which are not visible. In fact, the sun radiates significant amounts of power over a range of wavelengths much wider than the visible band. The way the total radiated power is distributed over the wavelengths emitted is shown in the uppermost curve in Figure 32. The sun's radiation is very nearly that of a black body at a temperature of 6000 K. The other curves in Figure 32 show the distribution of radiated power from black bodies at several lower temperatures.

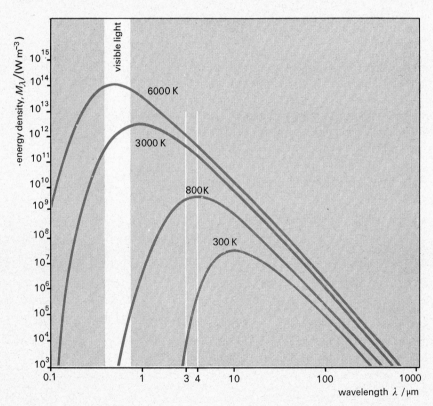

Figure 32 The distribution of power radiated by a black body at different temperatures

The differences between the power distributions at different temperatures are exploited by radiation pyrometers, so let us be clear as to what these curves mean. Let us consider the 6000 K curve. It describes the radiation emitted by one metre squared of the surface of a black body at that temperature. The power of the radiation emitted in any band of wavelengths is given by the value of the curve in that band multiplied by the width of the band. Another way of looking at it is that the power in any band equals the area under the curve in that band. Take, for example, the visible radiation emitted by a black body at 6000 K. The power equals the area of the shaded region in Figure 32, or approximately 10^{14} W m^{-3} times the width of the band, 0.35 µm, or about 35 MW per metre squared of surface area.

37

How would you find the total power radiated by a black body (per metre squared of its surface area) at *all* wavelengths?

By finding the total area under the curve.

By examining the way in which the area under the curves in Figure 32 increases, you can see that the power emitted by a black body increases rapidly with increasing temperature. In fact the total power increases in proportion to the fourth power of the thermodynamic temperature of a black body. This relationship is called *Stefan's law*

$$M = \sigma T^4,$$

where σ is a constant called *Stefan's constant* and has the value

$$\sigma = 5.67 \times 10^{-8} \text{ W m}^{-2} \text{ K}^{-4}.$$

Stefan's law

Stefan's constant

SAQ 7

Assume that the filament of a 60 W light bulb is at a temperature of 3000 K and that its radiation approximates to black-body radiation. What is the surface area of the filament? How much of the 60 W is in the visible band?

SAQ 7

Some radiation pyrometers measure radiation at all wavelengths. Their response is thus described by Stefan's law.

Other radiation pyrometers only measure radiation in a narrow wavelength band, with the band carefully chosen to suit the application. (I shall explain why this is done in the next section.) Because the curves in Figure 32 are nested one within the other, the power radiated at any wavelength always increases with increasing temperature. (Consider, for example, how the power increases with increasing temperature along the line drawn at 3 μm.) To describe the way the power varies with temperature quantitatively we need the algebraic expression which the curves in Figure 32 represent.

It is called *Planck's law* and is given below for reference only.

Planck's law

$$M_\lambda = \frac{c_1}{\lambda^5 \{\exp(c_2/\lambda T) - 1\}},$$

where λ is the wavelength, $c_1 = 3.74 \times 10^{-16}$ W m^{-2}, $c_2 = 1.44 \times 10^{-2}$ m K.

There is a third method used in some radiation pyrometers. This is to compare the relative amounts of power at two different wavelengths. If you compare the power at 3 μm and 4 μm in Figure 32, you will see that the relative values are different on each of the different temperature curves.

5.3 Radiation from objects other than black bodies

The principles in the last section describe how to relate radiated power to temperature for a black body. Unfortunately, there is no way to guarantee that the object whose temperature you want to measure will turn out to radiate like a black body. In fact, a black body, in addition to being the best absorber of electromagnetic radiation, is also the best emitter. Any other object at the same temperature will emit less power or at best the same power at any wavelength. (See Figure 33 for an illustration of this.)

The reason for this is that, as I have said, when any object is at the same temperature as its surroundings its absorption and radiation of energy

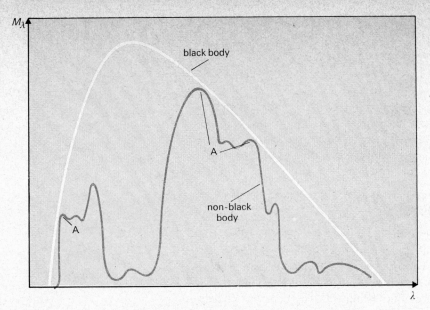

Figure 33   Black-body radiation compared with radiation from a possible non-black body

must balance. Since a non-black body absorbs less than a black body would in the same situation, it must also radiate less. This principle applies not only to the total radiation, but to the radiation at each wavelength. It can be summarized as *Kirchhoff's law*.

Kirchhoff's law. If an object at a certain temperature T absorbs a certain fraction ε of the radiation incident upon it at a wavelength λ, then it will emit the same fraction ε of the radiation at that wavelength that a black body at temperature T would emit.

The fraction ε is called the *emissivity* of the object. As in the example in Figure 33, it may vary with the wavelength. It may also vary with the temperature of the object.

Kirchhoff's law

emissivity

Figure 34   Kirchhoff's law. The upper figure absorbs half the incident radiation. (The lower black body absorbs it all.) The upper figure thus emits half what the black body emits

What is the emissivity of a black body? Does it vary with temperature and wavelength?

You can now see that it is not sufficient to measure the radiation emitted by an object to determine its temperature. You must also know its emissivity. Emissivities are not always easy to determine and thus constitute one of the major sources of uncertainty in radiation pyrometry. There are several ways of overcoming this problem:

1 The emissivity may have been measured and reported in the technical literature. (This is the easy solution!)

2 It may be possible to calibrate the radiation pyrometer on the actual object being tested by using some other transducer such as a thermocouple or resistance thermometer. Of course, if you are using a radiation pyrometer because the nature of the measurement makes contact impossible (e.g. the interior of a blast furnace), this technique will not be of much help.

3 The object may have an emissivity which is nearly 1 at certain wavelengths (see for example, the points marked A on Figure 33). If the radiation pyrometer has a filter in the path of the radiation which blocks all radiation except that in a band where the target has an emissivity of nearly 1, the target will appear to be a black body to the pyrometer. This is a very widely used technique. A wide variety of filters is available to suit the specialized needs of different industrial applications.

4 If it is possible to attach a device such as the one in Figure 35 to the target, the radiation emitted by the target, when viewed through the opening in the device, will approximate that of a black body. The device is a cavity with reflecting walls and an opening which is small compared to its interior surface area. Radiation entering the cavity will either be absorbed by or reflected off the target or reflected off the walls of the cavity. Each time it is reflected off the cavity and back to the target, some will be absorbed. After multiple reflections the fraction not absorbed by the target will be very small. In this way the target can be made to appear to be a black body with an effective emissivity of about 0.99.

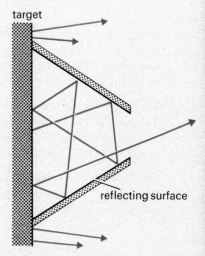

Figure 35 *When the target is viewed through the opening in this cavity it appears like a black body*

Incidentally, by a similar argument, the radiation emitted from the interior of an oven with a small hole in it also approximates that of a black body when viewed through the hole. Such furnaces are used as reference sources to calibrate radiation pyrometers.

The emissivity of the target itself is not the only emissivity problem encountered in radiation pyrometry. The radiation is transmitted from the target to the pyrometer through air.

Is the emissivity of air nearly unity or nearly zero in the visible band?

The emissivity is nearly zero. Since it transmits most of the light in that band, very little is left to be absorbed. Remember, by Kirchhoff's law, the fraction of radiation absorbed at any wavelength is equal to the emissivity at that wavelength. Thus the emissivity of air in the visible band is nearly zero.

However, air is not so transparent to all radiation. The fraction it absorbs rises considerably at some wavelengths in the infrared (see Figure 36) so, at those wavelengths, air is relatively opaque. Since the radiation from a target must pass through air on its way to a radiation pyrometer, the absorption characteristics of the atmosphere will affect the radiation received by the pyrometer. Moreover, the atmospheric absorption depends

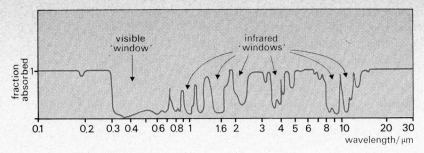

Figure 36 Atmospheric absorption as a function of wavelength

upon the distance between the target and the pyrometer. By using filters to limit the radiation received by the pyrometer to a band in one of the atmospheric 'windows' where the absorption is relatively low, this problem can be largely eliminated.

Similarly, absorption by any lenses used in the pyrometer's optical system also modifies the radiation its detector receives. However, as this is known in advance, the calibration of the pyrometer can be made to include its effects.

5.4   Pyrometer optical systems and detectors

The radiation laws we have considered up to now give the total radiation in all directions from a radiating object. Of course, when a radiation pyrometer is aimed at a target it only intercepts a fraction of the emitted radiation. Figure 37 shows a simplified diagram of a pyrometer in which the 'optical system' is simply a hole, or aperture. When the sensing head of the pyrometer is in position 1, the only part of the radiation from the target which can reach the detector is that emitted by area A–A' and lying within

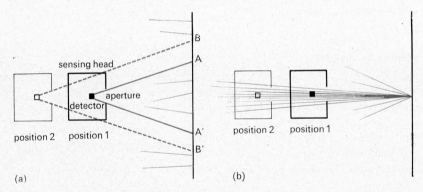

Figure 37   The distance from the target to the detector does not affect the reading so long as the target fills the cone of acceptance of the aperture

the cone of acceptance of the aperture. If the sensing head is moved to position 2, emission from B–B', a larger area, can reach the detector. On the other hand, as shown by Figure 37(b), a smaller fraction of the radiation emitted by each point on the target reaches the detector in position 2 than in position 1. The fraction received is proportional to the square of the distance between the sensing head and the target. The area of the target within the cone of acceptance of the aperture is also proportional to the square of the distance. Thus the decrease in radiation received from each point on the target is just balanced by the increased area when the sensing head is moved from position 1 to position 2. The energy reaching the detector thus does not depend upon the distance to the target so long as the target completely fills the cone of acceptance of the aperture.

If a lens is used to focus the incoming radiation (as shown in Figure 38) instead of just an aperture, the effect is to decrease the area of the target seen by the detector for a given aperture size and distance. The area seen without the lens can be seen from the dashed lines in the figure. If the area was decreased by making the aperture smaller rather than using a lens, the amount of radiation reaching the detector would also be decreased. Thus, for a given target size, the use of a lens increases the sensitivity of the radiation pyrometer.

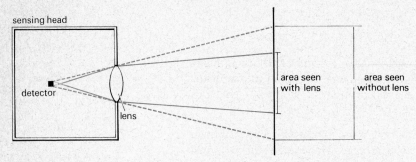

Figure 38   The effect of a lens in the optical system is to increase the effective aperture

Most of the types of temperature transducer described earlier in this unit are used as detectors in radiation pyrometers. Some detectors use thermocouples. Either a single thermocouple or several connected in series to form a *thermopile* are used (see Figure 39). The output voltage of a thermopile consisting of n thermocouples with all hot junctions at the same temperature and all cold junctions at the same temperature is just n times that of a single thermocouple.

thermopile

Some radiation pyrometers use thin flakes or films of metals or semiconducting material as resistance temperature sensors. Reistance sensors of this form are called *bolometers*.

bolometer

Some pyrometers use detectors which are not thermal detectors, but instead measure the radiation directly using the photoelectric effect.

5.5   Transducer example 5: A radiation pyrometer

Let us consider in some more detail the operation of the infrared radiation thermometer (Figure 30) with which we began this section. The diagram in Figure 40 shows the operation of the optical system. Some of the in-coming radiation passes through the beam-splitting mirror and through the eyepiece so that the operator can aim the device and focus its objective lens. Most of the incident radiation, however, is reflected off a

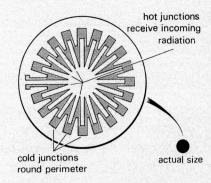

Figure 39   A thermopile radiation detector

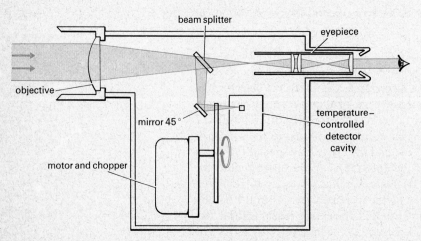

Figure 40   Schematic diagram of an infrared radiation thermometer

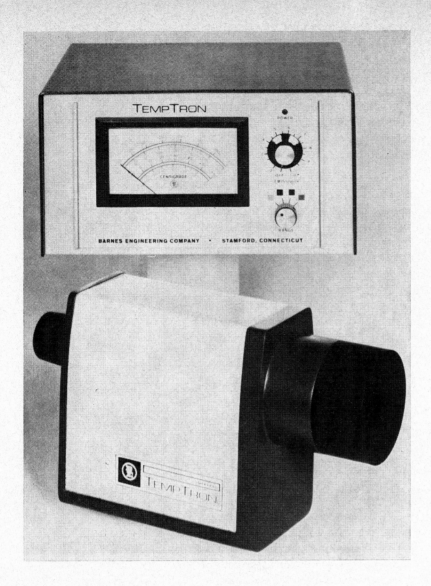

Figure 30   (repeat) TempTron infrared radiation thermometer

SPECIFICATION

TEMPTRON INFRARED THERMOMETER MODEL IT-7

Spectral band/μm	Temperature ranges/°C
0.7–1.0	500–1000 in four ranges
	1000–3000 in four ranges
1.9–2.6	100– 520 in four ranges
	500–2500 in four ranges
	30– 400
2.3–3.9	50– 650
5.0–6.2	50– 500
	200–1700

Field of view
0.7 °, 0.35 ° or short focus.

Accuracy
1% of span on most ranges.

Sensitivity and repeatability
½% of span.

Response time
50 ms (10%–90%).

Emissivity adjustment
0.2–1.0.

Allowable ambient temperature
0 °C–40 °C.

Power input
115°230 V, 50 or 60 Hz, 50 W maximum.

Dimensions (approximate)
Sensing head (size) H – 15.51 cm,
                       W – 8.89 cm,
                       L – 30.5 cm,
            (weight) 3.63 kg,
Indicator unit (size) H – 13.34 cm,
                       W – 26.04 cm,
                       T – 24.13 cm,
            (weight) 3.96 kg.

second mirror to the detector, which is located at the end of a temperature-controlled cavity. At the entrance to the temperature-controlled cavity the radiation is periodically interrupted by the rotating vanes of the *chopper*. The vanes have a highly reflective gold plating so that when they block the cavity entrance they also reflect back any radiation from within the cavity. The material of the detector, plus a set of optical filters at the end of the cavity, determine the band of wavelengths and temperature range of this infrared thermometer. For most of the ranges, thermistors are used, although the semiconductor used in them is different for different ranges.

chopper

The function of the cavity is to provide a known environment for the detector. The temperature of the cavity is kept at 45 °C±0.1 °C. If there is no in-coming radiation (as, for example, when the cavity entrance is blocked by the chopper), the only radiation reaching the detector will be from the cavity. The detector will reach an equilibrium when the total radiation it receives equals the total radiation it emits. This will occur when its temperature is equal to that of the cavity. When the radiation from the source, whose temperature is greater than 45 °C, is incident on the detector (when the cavity entrance is not blocked by the chopper) the total radiation it receives will be greater. Thus the detector will reach equilibrium at a temperature higher than that of the cavity.

SAQ 9
SAQ 9

How will the description above be changed if the temperature being measured is lower than the temperature of the cavity?

Now we can see that the effect of the chopper is to cause the detector temperature to rise and fall alternately. The detector has a response time of about 1 ms. The chopper interrupts the in-coming radiation at a rate of about 100 Hz. That is, the window is alternately open for 5 ms and then closed for 5 ms. Thus the detector has time to reach within 1 per cent of equilibrium each time the cavity entrance is opened or closed, as shown in Figure 41.

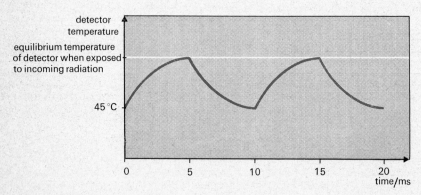

Figure 41   The temperature of the detector against time

The thermistor detector forms one arm of a bridge. The bridge output is a voltage waveform of the same form as the temperature waveform in Figure 41. Its amplitude depends upon the temperature being measured.

The bridge output voltage is amplified, converted to d.c. and fed to a meter.

The purpose of the chopper and the resulting a.c. waveform is to improve the accuracy and stability of the measurements.

Section 6

Interlude: The response time of first-order transducers

Suppose you are using a transducer to measure some physical quantity which changes with time. The transducer produces an output signal, electrical or otherwise, which represents the measured quantity. However, no matter what kind of transducer it is, the value of the output signal at any instant of time will represent the value of the measured quantity *not* at that instant but at some earlier instant of time. There will always be some delay between the occurrence of an event and the transducer signal representing it.

I have mentioned this in passing at several points in this unit. In this section we shall look at the problem in more detail, but let us first review what was described in transducer example 2.

Figure 16 shows how the thermocouple voltage changed after being immersed in water at 90 °C. The thermocouple was immersed in the water at the time shown by the immersion marker pulse. It did not reach its final steady-state output level (corresponding to 90 °C) until some time later. The basic reason for this delay is that it takes time for the measuring junction to reach the temperature of the surrounding water.

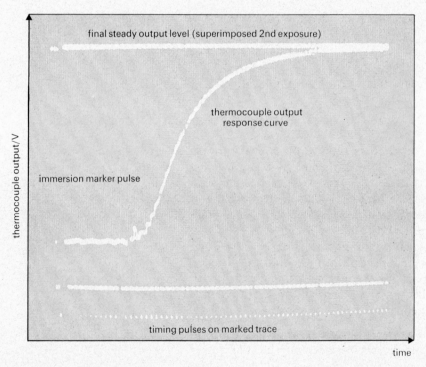

Figure 16 (repeat) An oscilloscope trace showing the response of a thermocouple probe to a rapid immersion in water at 90 °C

The time the thermocouple voltage took to get to about 63 per cent of its final value was called the *time constant*, and was listed in Table 3 for various probe diameters. It ranged from 0.15 s to 3.90 s. The thermocouple output did not reach its final steady-state output voltage until a considerably longer time had passed.

time constant

If these thermocouples are to be used in an application in which the temperature changes over a period of minutes or hours, the delays in the reading due to the thermocouple response will be of no concern. On the other hand, these thermocouples clearly cannot be used for applications in which temperature changes over fractions of a second must be recorded.

To explain in more detail the way in which the thermocouple and other temperature transducers respond to changing temperatures, and in particular, the concept of time constant, I am going to represent their behaviour by a simplified model of it. I am going to describe temperature transducers as *first-order linear systems*, a name I shall explain shortly. In Units 8/9/10 you will meet a second simplified model, a second-order linear system, which is useful for describing the way a different group of transducers respond to changes in the variable they are measuring.

first-order linear system

I shall restrict the discussion to sudden, or 'step', changes of temperature, as shown in Figure 42. The sudden immersion of the thermocouple into hot water approximated such a step change.

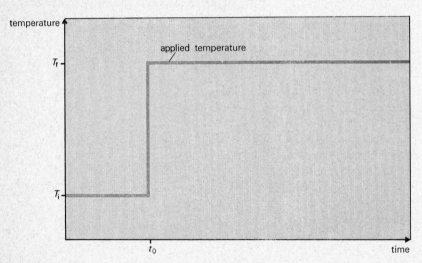

Figure 42   A step change of temperature from T_i to T_f occurring at time t_0.

Let us assume that before time t_0, the thermocouple output voltage corresponds to the initial temperature T_i the thermocouple is measuring. After t_0, for some time there will be a difference between the temperature the thermocouple is measuring (which is now T_f, the final temperature) and the temperature corresponding to its output voltage. The basic assumption upon which the following discussion depends is that the temperature of the thermocouple measuring junction will change towards the temperature it is measuring at a rate proportional to how different it is from that temperature. The greater the temperature difference between the water and the thermocouple junction, the faster heat flows into the junction and the faster its temperature rises. Another assumption in my discussion is that there is negligible delay between the establishment of a temperature change at the junction and the corresponding output voltage.

The discussion following will be clearer if I compare the applied temperature to the temperature of the thermocouple measuring junction rather than to the corresponding thermocouple output voltage, so I will do that from now on.

Let us see how the basic assumption helps to explain the way the temperature of the thermocouple measuring junction changes. As you can see in Figure 43, the rate of rise of this temperature is greatest just after t_0, when it is furthest from T_f. At time t_1, when the temperature is half-way to T_f, its rate of rise is half that just after t_0. When the temperature reaches three-quarters of the way to T_f, at time t_2, its rate of rise is only one-quarter the rate just after t_0.

> Will it take longer for the temperature to get half-way to T_f or to get from there to three-quarters of the way to T_f?

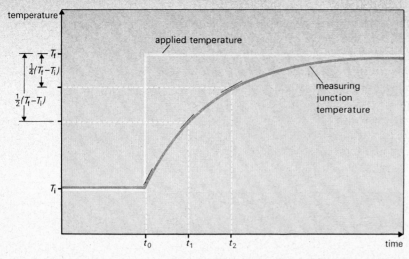

Figure 43   The slope (i.e. the rate of change) of the curve is half at t_1 what it is at t_0, and half again at t_2

Both will take the same length of time, because the latter change takes place at half the rate and so will cover half the temperature interval of the first in the same length of time. In fact, no matter what the temperature of the measuring junction is, the time interval until it is only half as far from T_f will be the same. This is illustrated in Figure 44.

The following self-assessment question illustrates a major use for the principles I have just been discussing.

SAQ 10

SAQ 10

Suppose the time intervals marked A, B and C in Figure 44 are one second each. Estimate how long after t_0 the measuring junction temperature will pass 75 °C.

The reasoning I used to conclude that the time intervals A, B and C in Figure 44 were the same applies equally well where the fractional change in temperature difference is other than a half. For example, if the temperature difference at the end of intervals A, B and C had been one-third what it was at the beginning, the three intervals would still be equal. We are now ready to discuss the time constant.

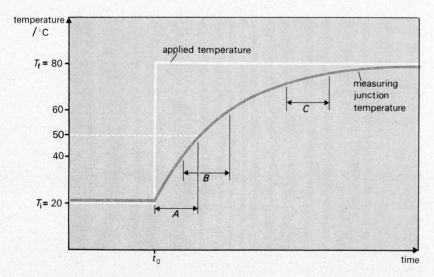

Figure 44   At the end of each time interval A, B and C the temperature is half as far from T_f as it is at the beginning

47

I have described the time constant as the interval in which the temperature change covers about 63 per cent of the difference between its initial and final temperatures. More precisely, if, at the end of a time interval τ, the difference between the applied temperature T_f and the measuring junction temperature is $1/e$ what it was at the beginning of that interval, then τ is the time constant of the thermocouple. The number e is discussed in 'Mathematics for Instrumentation', section 4.2. The value of e is e ≈ 2.7183 approximated to four decimal places. Thus $1/e \approx 0.37$. If a temperature difference is reduced by a factor of about 0.37, it has changed by a factor of about $1.00 - 0.37 = 0.63$, or 63 per cent (see Figure 45).

time constant defined

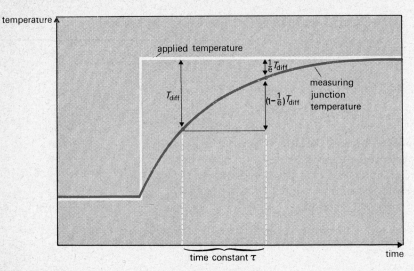

Figure 45   An illustration of the definition of time constant

After two intervals of one time constant each the temperature difference will be reduced by a factor of $(1/e)(1/e) = 1/e^2$, because in the second interval it will be reduced by a further factor of $1/e$.

After an interval of three time constants it will be reduced by a factor of $1/e^3$, and so on. This is summarized in Table 10.

Table 10

At the end of a time interval of length	The temperature difference is reduced by a factor of	The percentage of the difference covered in that interval is
τ	$1/e \approx 0.368$	63.2%
2τ	$1/e^2 \approx 0.135$	86.5%
3τ	$1/e^3 \approx 0.050$	95.0%
4τ	$1/e^4 \approx 0.018$	98.2%
5τ	$1/e^5 \approx 0.007$	99.3%

SAQ 11

SAQ 11

Suppose you take a thermocouple from water at 70 °C and place it in water at 20 °C. If the thermocouple output voltage must correspond to a temperature of 20 ± 1 °C within 8 s, estimate the maximum time constant the thermocouple may have.

Now I am going to relate the time constant to the basic assumption that I made earlier, and also explain how this assumption ties in with the name 'first-order linear system'.

The basic assumption that the rate at which the temperature of the thermo-couple measuring junction T changes is proportional to the difference between T and the final value of the applied temperature T_f can be written algebraically as

$$\frac{dT}{dt} \propto T_f - T$$

or $\quad \dfrac{dT}{dt} = k(T_f - T),$ (1)

where k is the constant of proportionality. This expression is called a *first-order linear differential equation*. Such equations are discussed in section 4.5 of 'Mathematics for Instrumentation'.* Any system described by a first-order linear differential equation is called a first-order linear system. The description I have given of the response of a thermocouple to a step change in applied temperature in fact also describes the response of *any* first-order linear system to a step change in its input. In particular, it quite accurately describes the response of all the temperature transducers in Unit 3 as well as some transducers you will meet in other parts of the course.

It turns out that the proportionality constant k in equation (1) is the reciprocal of the time constant τ. Thus equation (1) can be written as

$$\frac{dT}{dt} = \frac{1}{\tau}(T_f - T).$$ (2)

This agrees with what I have said up to now: the larger τ is, the slower is the rate of change of the temperature of the measuring junction.

To find the way in which T changes we need the solution to equation (2). To find the solution we need the value of T at the starting time t_0. That value is T_i. Let us also say that $t_0 = 0$. The solution is then

$$T = T_f - (T_f - T_i)\exp(-t/\tau).$$ (3)

(Note that $\exp(-t/\tau)$ means $e^{-t/\tau}$.) Equation (3) is the algebraic expression for the measuring-junction temperature curves in Figures 43–5. It says that, at any moment, T equals the final temperature T_f less a term which represents the difference between T_f and the value of T at the moment. At the initial time, $t = 0$, and so this difference term equals $T_f - T_i$, because $\exp(0/\tau) = 1$. Thus

$$T = T_f - (T_f - T_i) = T_i.$$

Then, as t increases, $\exp(-t/\tau)$ gets smaller and smaller so that the difference term in equation (3) gets smaller. That is, T gets closer and closer to T_f.

Let us go over equation (3) again, but in terms of the numerical values in Figure 44. I have repeated Figure 44 in Figure 46 with a time scale added. Thus $T_f = 80\ °C$, $T_i = 20\ °C$ and their difference is $T_f - T_i = 60\ °C$. The time at which the temperature of the measuring junction reaches 20 °C plus 63 per cent of the 60 °C difference is the time constant τ.

$$0.63 \times 60 = 39,$$

so 63 per cent of the way between 20 °C and 80 °C is 20 °C + 39 °C = 59 °C.

From the curve, the time at which $T = 59\ °C$ is 1.4 s after t_0, so $\tau = 1.4$ s.

(Alternatively, τ can be calculated from my earlier statement that intervals A, B and C are 1 s.)

* *'Mathematics for Instrumentation' only explicitly describes this equation for the case where $T_f = 0$.*

49

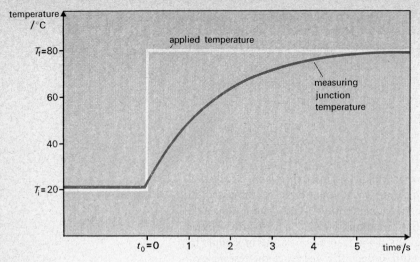

temperature /°C

$T_f = 80$

60

40

$T_i = 20$

applied temperature

measuring junction temperature

$t_0 = 0$ 1 2 3 4 5 time/s

Figure 46

Using these values, equation (3) becomes

$$T = \{80 - 60 \exp(-t/1.4 \text{ s})\} \; °C. \tag{4}$$

At $t = 0$ (i.e. at t_0), equation (4) gives

$$T = \{80 - 60 \exp(0)\} \; °C = \{80 - 60(1)\} \; °C = 20 \; °C.$$

At a time equal to the time constant, $t = 1.4$ s, equation (3) gives

$$T = \{80 - 60 \exp(-1.4/1.4)\} \; °C = \{80 - 60 \exp(-1)\} \; °C$$
$$= \{80 - 60(1/e)\} \; °C.$$

The difference term has been reduced by a factor of $1/e$ during the interval from 0 to 1.4 s.

At a time equal to twice the time constant, $t = 2.8$ s, equation (3) gives

$$T = \{80 - 60 \exp(-2.8/1.4)\} \; °C = \{80 - 60 \exp(-2)\} \; °C$$
$$= \{80 - 60 \, (1/e^2)\} \; °C.$$

The difference term has now been reduced by a further factor of $1/e$. We are confirming that the properties of the time constant which I described previously do in fact follow from equation (3) as well.

In summary, after any interval of time equal to the time constant, the difference term is reduced by a further factor of $1/e$, that is, the temperature is closer to its final value by a factor of $1/e$ than it was at the beginning of that interval.

Section 7

Summary

In this unit I have described some of the more important principles and transducers involved in measuring temperature.

Before attempting to measure temperature, it is important to know what temperature means, and which *temperature scales* are commonly used. The *thermodynamic scale*, with an interval of one *kelvin* (K) is described. It is defined by two points: *absolute zero* and the *triple point of water*. Its relation to the *Celsius* (or *centigrade*) *scale* is discussed. The practical realization of the thermodynamic scale in the *International Practical Temperature Scale* is shown in Table 1.   Section 1

Temperature can be measured as changes in volume, length or pressure. Concepts involved are: *coefficient of linear expansion* (bimetallic strip), *differential expansion* (liquid-in-glass thermometer; bulb, capillary and pressure-sensor type). Response time is also discussed.   Section 2

Thermocouple (*thermoelectric*) transducers utilize the *thermoelectric effect* (*Seebeck effect*). The junction of two dissimilar metals has a *contact potential*, which varies with temperature.   Section 3 / 3.1

The voltage measured by a meter in a thermocouple circuit depends on the difference between the temperatures at the measuring junction and the *reference junction*. For greatest accuracy the reference-junction temperature must be measured accurately. It is common practice to hold it at a known, fixed, temperature. *Extension wires* can be used to avoid long and expensive thermocouple wires. In an example of a *thermocouple electronic thermometer*, the reference junction is at ambient temperature and is compensated for electronically. The characteristics of various thermocouple materials are described. The construction and application of a range of thermocouple probes are discussed.   3.2 / 3.3 / 3.4 / 3.5

Resistive temperature transducers change their electric resistance with temperature. Both metallic resistors and semiconductors are used. The resistance of a metallic resistor increases approximately linearly with temperature, $R = R_0(1 + aT)$. The proportionality constant a is called the *temperature coefficient of resistance*.   Section 4 / 4.1

Metallic resistance probes are constructed in various forms, depending on the nature of the material whose temperature is to be measured.   4.2

The specification of a *platinum resistance temperature sensor* is discussed.   4.3

Semiconductor temperature transducers are called *thermistors*. They have a non-linear (exponential) resistance change with temperature,   4.4

$$R = A \exp(B/T),$$

with a temperature coefficient which is *negative*, and is dependent on the temperature. Over a substantial part of the range the coefficient is greater than that of the metallic resistors, so thermistors are more sensitive.

A common measuring circuit for resistive temperature transducers is the *bridge circuit*. Because of the relatively large fractional resistance changes occurring, non-linearity occurs in the bridge circuit.   4.5

A proprietary *electronic thermometer* uses a thermistor in a bridge circuit. The non-linearities in the thermistor and the bridge tend to cancel one another and good overall linearity is achieved.   4.6

Radiation pyrometers measure the *infrared radiation* emitted by hot bodies. Section 5
The radiation is used to heat a detector whose temperature is related to
that of the hot body. *Infrared* radiation has a *wavelength* and *frequency* 5.1
just beyond the red end of visible spectrum.

Black bodies absorb all incident energy. 5.2

The power radiated by a black body is determined by Stefan's law,
$M = \sigma T^4$, where σ is *Stefan's constant* $= 5.67 \times 10^{-8}$ W m^{-2} K^{-4}. A major 5.3
source of uncertainty in radiation pyrometry is the *emissivity* ε of the object
whose temperature is being measured. The emissivity is the ratio of the
radiation emitted by the object to that emitted by a black body at the same
temperature. *Kirchhoff's law* states that the same ratio applies to the
radiation *absorbed* by the object as compared to a black body (which
absorbs all radiation). The absorption of the air and the optical system
are also important in making measurements with a radiation pyrometer.
The useful property of independence of measuring distance is possessed 5.4
by readings from radiation pyrometers. Various temperature detectors are
used to measure the radiation falling on the pyrometer target, including
thermopiles and *bolometers*.

A proprietary radiation pyrometer is discussed. It uses a *chopper* which 5.5
periodically interrupts the radiation entering the temperature-controlled
cavity where the detector is located. Accuracy and stability are improved
with this arrangement.

The *response time* of a *first-order linear system* is discussed with reference Section 6
to temperature transducers. It is an *exponential* function of temperature
describing the delay which arises in temperature measurements.

Self-assessment answers and comments

SAQ 1

The net contact potentials at both the hot and cold junctions will be the same in both cases. That is, $V_A = V_6 - V_5$ and $V_B - V_C = V_2$. The voltmeter reading will thus have the same magnitude in both cases. (The signs of the two readings may or may not be the same, depending upon which way round the voltmeters are connected.)

SAQ 2

From the table, the thermocouple voltage at 100 °C is 6.32 mV with a reference temperature of 0 °C. Since the actual voltage is 5.13 mV, the reference temperature must have been greater than 0 °C. The difference between the two voltages is 1.19 mV, which corresponds to a reference temperature of 20 °C according to the table. Adding 1.19 mV to 3.15 mV gives 4.34 mV, which is half way between the table entries for 60 °C and 80 °C. Thus it represents a temperature of about 70 °C.

SAQ 3

Using the formula $R = R_0(1 + aT)$, with $a = 0.0039$, $R_0 = 100\ \Omega$, and $T = 50$ °C, gives $R = 119.5\ \Omega$.

SAQ 4

Using the formula $R = R_0 \exp\{B(1/T - 1/T_0)\}$, with $R_0 = 100\ k\Omega$, $T_0 = 300$ K, $T = 600$ K and $B = 3000$ K gives

$$R = 10^5 \exp\left[3000\left(\frac{1}{600} - \frac{1}{300}\right)\right] \Omega$$

$$= 10^5 \exp\left[3000 \times \frac{-1}{600}\right] \Omega$$

$$= 10^5 \exp(-5)\ \Omega$$

$$= 670\ \Omega.$$

SAQ 5

The effect of the resistance of the leads depends upon where the connections to the voltmeter are made. If they are made very near the temperature transducer, the voltage measured will be independent of the resistance of the leads to the constant current source. This is because the current through the transducer is fixed by the current source and is not affected by changes in the lead resistance. This is an important feature of this circuit. The resistance of the leads to the voltmeter have a negligible effect if it is much less than the voltmeter resistance.

SAQ 6

In the equation

$$V_o = \left[\frac{R_T}{R_T + R_2} - \frac{R_3}{R_3 + R_1}\right] V_s$$

the non-linearity arises because of the R_T in the denominator of the first term. Thus the smaller it is by comparison to R_2, the more linear will be the equation. However, the voltage across the resistive transducer is $V_s R_T/(R_T + R_2)$, so increasing R_2

relative to R_T will decrease the fraction of V_s appearing across R_T. Thus changes in R_T will produce a proportionately smaller voltage across the bridge output terminals.

SAQ 7

If the temperature of the filament is 3000 K, the power it will emit (per metre squared of surface area) is given by Stefan's law as

$$M = 5.67 \times 10^{-8} \times (3000)^4\ \text{W m}^{-2}$$

$$= 4.6\ \text{MW m}^{-2}.$$

Assuming that all the 60 W is radiated, the surface area will be

$$A = \frac{60\ \text{W}}{4.6\ \text{MW m}^{-2}}$$

$$= 13 \times 10^{-6}\ \text{m}^2$$

$$= 13\ \text{mm}^2.$$

To find the fraction of the 60 W emitted in the visible band we can perform an approximate numerical calculation of the white area in Figure 32 under the 3000 K curve. Taking 10^{12} W m^{-3} as an average value of the curve and 0.35 μm as the width of the band gives

$$M_\lambda = 0.35 \times 10^6\ \text{W m}^{-2}$$

$$= 0.35\ \text{MW m}^{-2}.$$

Thus, of the 60 W going into the bulb, 0.35/4.6, or 7.5 per cent, ends up as visible light.

SAQ 8

The emissivity of a black body is unity at all temperatures and wavelengths.

SAQ 9

Because the target is at a lower temperature than the cavity, less radiation will reach the detector when the cavity is open than when it is closed. Thus the detector temperature will fall below the cavity temperature when the cavity is open and will rise towards the cavity temperature when the cavity is closed.

SAQ 10

The initial difference between the temperature of the thermocouple measuring junction and the applied temperature is 60 °C. After one second this difference is halved to 30 °C, after a further second the difference is halved again to 15 °C. Continuing in this way we can conclude that the difference will become less than 5 °C between 3 and 4 seconds after t_0.

SAQ 11

The total temperature change for the thermocouple is 50 °C. The 1 °C tolerance allowed after 8 s is 2 per cent of that change. According to Table 10 this will not occur until a time interval of 4τ has passed. Thus 4τ must be less than 8 s, or the maximum value of the time constant is about 2 s.

Unit 4

Contents

1 Sensing elements 3
1.1 Some basic principles 3
1.2 Sensing elements for force transducers 4
1.3 Transducer example 1: A strain-gauge load cell 5
1.4 Sensing elements for torque transducers 8
1.5 Sensing elements for pressure transducers 10

2 Resistive potential dividers (pots) 12
2.1 Linear-displacement pots 12
2.2 Other displacement pots 14
2.3 Transducer example 2: A pressure transducer using a potential divider 15

3 Magnetic transducers 17
3.1 Introduction 17
3.2 The differential transformer 17
3.3 The phase-sensitive detector 19
3.4 Transducer example 3: A differential transformer for measuring displacement 21
3.5 Variable-inductance (variable-reluctance) displacement transducers 23
3.6 Transducer example 4: An inductance-bridge pressure transducer 24

4 Capacitive transducers 27
4.1 Capacitors and capacitance 27
4.2 Capacitance variation caused by displacement 27
4.3 Capacitance-measuring circuits 29
4.4 Transducer example 5: A capacitive differential pressure transducer 34

5 Piezoelectric transducers 35
5.1 Piezoelectricity 35
5.2 Piezoelectric transducers 35
5.3 The step response of a piezoelectric transducer 36
5.4 Measuring circuits 37
5.5 Charge measurement and charge amplifiers 38
5.6 Temperature effects 41
5.7 Transducer example 6: A quartz force link 41

6 Summary 44

Appendix   Units and conversion factors for pressure 46

Self-assessment answers and comments 47

Section 1

Sensing elements

1.1 Some basic principles

A familiar device for measuring force is the *spring balance* (Figure 1).
It measures the gravitational force on an object by applying that force to a
spring. The spring stretches by an amount which is in proportion to the
force, so that a scale measuring the deflection of a pointer at the end of the
spring can be calibrated in terms of the force applied.

Many transducers for measuring force and force-related quantities such as
torque and pressure use a similar principle. The force, pressure or torque
is applied to some form of elastic member which compresses, expands,
bends or twists. The resulting displacement or strain is then sensed by a
secondary transducer which converts it to an output signal. The output
signal can then be measured and calibrated in terms of force, torque or
pressure, as is appropriate.

For example, the pneumatic pressure transducer of Unit 1 uses a diaphragm capsule as its elastic member, and the displacement is represented
by a *pneumatic* output signal. All the transducers described in Unit 4
provide *electrical* output signals, but the same basic principle, of conversion to a displacement, or strain, and from that to an output signal, is
used. Thus many of the techniques used in displacement transducers and,
in fact, some actual displacement transducers are used as the secondary
transducers in force, torque or pressure transducers. That is why it is
convenient to discuss these four variables together. In this section I shall
describe sensing elements for force, torque and pressure. In the following
sections I shall describe transducers which can either be used as displacement transducers or as secondary transducers for force, torque or pressure.

Terms and units

Before we go any further, I should like you to make sure your basic ideas
on force, pressure and torque are correct by attempting the following
revision self-assessment questions:

spring balance

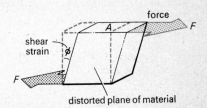

Figure 1   *A schematic representation of a spring balance*

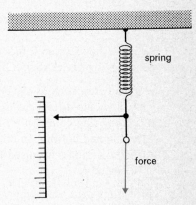

Figure 2

SAQ 1

State the relationship between *mass*, the *force* acting upon it, and its
resultant *acceleration*. State the SI unit of force.

SAQ 2

State the relationship between *pressure* and *force*, and state the SI unit
of pressure (defined in Units 1 and 2).

SAQ 3

Explain the terms 'shear stress', 'shear strain' and 'shear modulus',
and how they are related. (Use Figure 2 to help you.) State their SI
units. (These terms were defined in Unit 2.)

SAQ 4

What is meant by *absolute* pressure, *gauge* pressure and *differential*
pressure? (These terms were defined in Unit 1.)

SAQ 5

State the relationship between *head* and *pressure* (defined in Unit 1).

SAQ 1

SAQ 2

SAQ 3

SAQ 4

SAQ 5

SAQ 6

Explain what is meant by *torque*. (See Figure 3.) If this term is new to you, don't worry, just read on.

I shall start by discussing the sensing elements (the elastic members) which convert force, torque and pressure into displacement. They have a common principle of operation. When the applied force (or torque or pressure) is within the limits for which the sensing element has been designed, the deflection varies essentially linearly with the applied force (or torque or pressure). The shape and materials used depend upon the application.

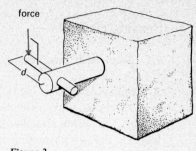

Figure 3

1.2 Sensing elements for force transducers

Figure 4 shows some force-sensing elements. When a load is applied to them they deflect as shown by the dashed lines.

In some cases, such as the *proof ring* of Figure 4(a), the displacement of one part of the member is measured with respect to that of another part. The proof ring commonly has a displacement transducer connected between top and bottom. In other cases, such as the solid cylinder of Figure 4(d), strain gauges are used to measure strain at carefully chosen points on the member.

proof ring

Proof rings were used originally, not as transducers, but as self-contained instruments for measuring force. The relative displacement of top and bottom is measured with a micrometer. Types for both compressive stress ('push') and tensile stress ('pull') are available. Their widespread use for many years has led to the improvement of their characteristics, and proof rings are now available with excellent linearity, temperature coefficient and freedom from *creep*. Creep is a permanent deformation of an elastic

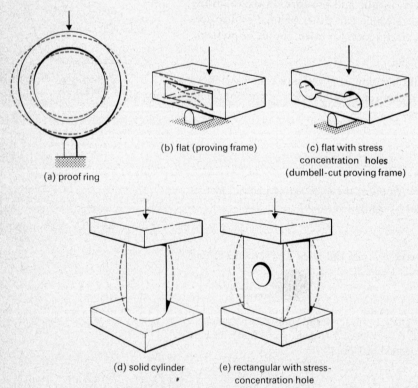

(b) flat (proving frame)

(c) flat with stress concentration holes (dumbell-cut proving frame)

(a) proof ring

(d) solid cylinder

(e) rectangular with stress-concentration hole

Figure 4   Some elastic sensors for force

member which has been under load for some time. When the load is removed, the member does not quite return to its original dimensions. It has acquired a permanent 'set' or zero error.

Because proof rings are so well-developed as force-measuring elements, they make an obvious choice as the elastic members of force transducers. The relative displacement between the top and bottom of the ring, when full load is applied, is commonly of the order of a millimetre. To measure this displacement one particular type of displacement transducer is a frequent choice. This type, the linear variable differential transformer (LVDT), is described in Section 3.

A second approach is to measure the strain in the proof ring, rather than its displacement. This is done in the transducer in the following example.

1.3 Transducer example 1: A strain-gauge load cell

In this example we shall look at a *load cell*: a force transducer used to measure weight. Figure 6 is a drawing of the complete load cell, while Figure 7 is an exploded view showing the construction of a typical model in this range.

load cell

Notice from Figure 7 that the force-sensing element used is either a ring-type or a column-type element. The type of element supplied depends upon the force range for which the transducer is supplied, with the ring-type element used for lower forces. By using different types of element the same full-range output can be obtained for different ranges.

The transducer described here is available for use in measuring tension only, compression only or as a universal load cell, measuring both tension and compression. To enable tension measurements to be made the load button (shown at the top of Figure 7) is removed, so that the load can be screwed directly into the sensing element. The only difference between the different types of transducer is in their rated accuracy. The deflection characteristics of the sensing element differ slightly in tension or compression. This can be accounted for in the calibration of the tension-only or compression-only models. However, since the universal model is used for both, its rated accuracy is slightly lower, because of the differences between tension and compression.

Each sensing element has six strain gauges mounted on it. To see why, consider the circuit diagram in Figure 5. Four of the strain gauges form the arms of the bridge. The other two are the resistors marked 'span

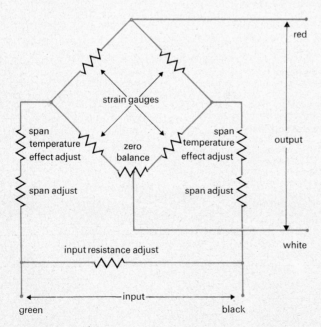

Figure 5   Circuit diagram for a load cell

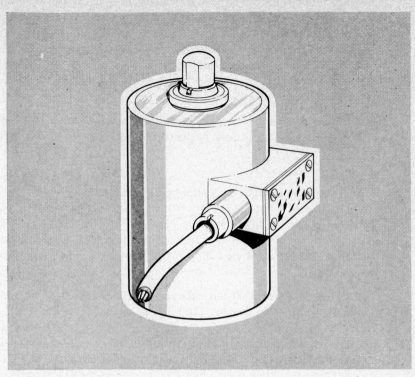

Figure 6   A BL4 type U2P1 strain-gauge load cell

SPECIFICATIONS
TABLE 1

	Tension only model	Compression only model	Universal model 89–134 N (20–30 lbf)	223–1340 N (50–300 lbf)
Performance				
Rated output (RO)—millivolts per volt	3		3	3
Calibration accuracy—% RO	0.1		0.25	0.25
Nonlinearity—% RO in tension	0.05	0.10	0.25	0.10
in compression	0.10	0.05	in both	in both
Hysteresis—% RO	0.05		0.10	0.05
Repeatability—% RO	0.05		0.10	0.05
Creep—% RO	0.05		0.10	0.05
Electrical				
Excitation, recommended	12 V, a.c. or d.c.		12 V, a.c. or d.c.	
maximum	20 V, a.c. or d.c.		20 V, a.c. or d.c.	
Zero balance—% RO	±1		±1	
Terminal resistance, input—ohms	350±3.5		350±3.5	
output—ohms	350±5.0		350±5.0	
Electrical connection	3 m cable		3 m cable	
Number of bridges	one		one	
Insulation resistance—bridge to ground	5000 MΩ		5000 MΩ	
shield to ground	2000 MΩ		2000 MΩ	
Temperature				
Temperature range, compensated	from −9 °C to 46 °C		from −9 °C to 46 °C	
safe	from −34 °C to 79 °C		from −34 °C to 79 °C	
Temperature effect on rated output	±0.0014% load/°C		±0.009% load/°C	
Temperature effect on zero balance	±0.0027% RO/°C		±0.0045% RO/°C	
Adverse Load Rating				
Safe overload—% rated capacity	500		300	500
Maximum sideload without damage— % rated capacity	10		10	10
Maximum bending moment without damage—newton metres	22.6		5.65	22.6
Maximum torque load without damage—newton metres	11.3		2.83	11.3

Mechanical data

Capacity/ N(lbf)	Deflection/ mm (in)	Weight/ kg (lb)	Natural frequency/Hz
89 (2)	0.36 (0.014)	1.1 (2.5)	280
137 (30)	0.30 (0.012)	1.1 (2.5)	400
223 (50)	0.33 (0.013)	2.3 (5)	240
446 (100)	0.28 (0.011)	2.3 (5)	350
891 (200)	0.23 (0.009)	2.3 (5)	550
1340 (300)	0.23 (0.009)	2.3 (5)	600

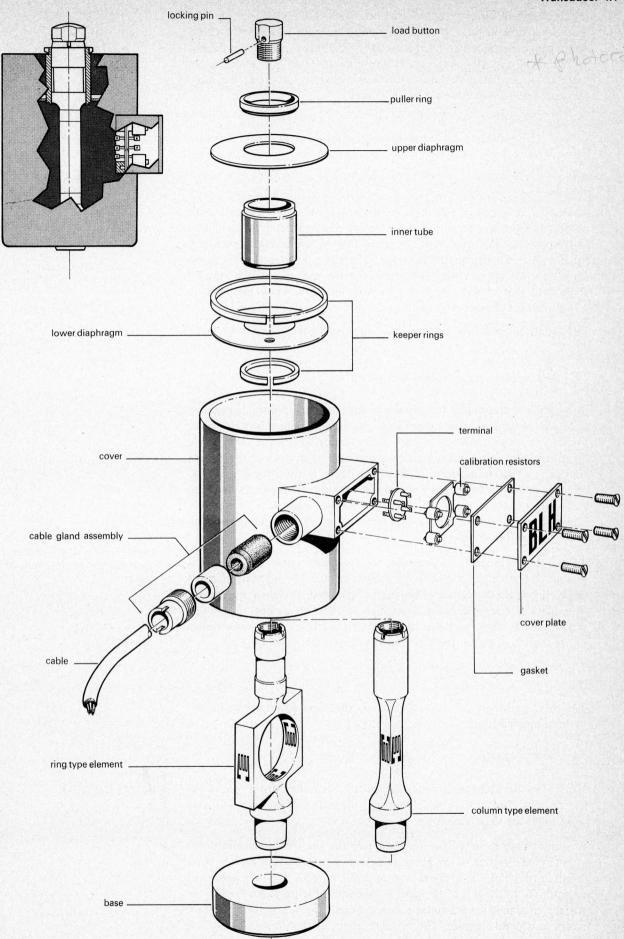

locking pin

load button

puller ring

upper diaphragm

inner tube

keeper rings

lower diaphragm

cover

terminal

calibration resistors

cable gland assembly

cover plate

cable

gasket

ring type element

column type element

base

Figure 7   Construction of a typical load cell

7

temperature effect adjust'. On the ring-type element the bridge resistors are arranged on the top, bottom and two sides of the inside of the ring. Two are visible in the exploded diagram. The 'span temperature effect' strain gauges are mounted on the outside of the ring-type element. One is visible in the exploded diagram. They are mounted so that they are not subjected to any strain (i.e. they are dummy gauges), but are at the same temperature as the four active strain gauges.

> The bridge provides temperature compensation because it has four active arms. Why then are the 'span temperature effect' strain gauges needed?

The use of a bridge of four active aims ensures that the bridge output accurately represents the strain in the sensing ring or shaft despite changes in temperature. However, it is not strain that we are trying to measure, but the force applied to the load cell. The strain produced for a given force depends upon Young's modulus for the material of the ring or shaft, and that changes with temperature too. The 'span temperature effect' resistors compensate for this change in Young's modulus.

The specifications for these load cells are given in Table 1.

1.4 Sensing elements for torque transducers

The choice of transducer for measuring torque is commonly determined primarily by quite a simple consideration: if two shafts transmitting the torque can be separated by a few centimetres or so, then a ready-made torque transducer can be inserted between them, so that the torque is then coupled via the transducer. If, however, there is insufficient space to insert a transducer, it is common practice to bond strain gauges onto the surface of the shaft, using the shaft itself as the elastic sensing element. The torque is sensed, in this case, in terms of the corresponding strain in the surface of the shaft.

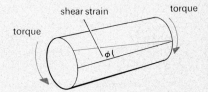

Figure 8 *The deformation and shear strain in a shaft resulting from an applied torque*

To use this technique it is necessary to know how much shear strain is present on the surface of a shaft for a given torque. The shear strain is the angle φ by which the surface of the shaft is distorted, as shown in Figure 8. (The black line represents the unstrained position of the line at an angle φ to it.) If the shear modulus of the shaft material is G and the radius of the shaft is a, the torque T is related to the shear strain by

$$T=\frac{\pi Ga^3\varphi}{2}.$$

The shear sensitivity can be considered as the shear strain in the surface per unit applied torque, or

$$\text{Shear sensitivity}=\frac{\varphi}{T}=\frac{2}{\pi Ga^3}.$$

Notice that the shear sensitivity is dependent upon the third power of the shaft radius. Thus a shaft which has half the radius of another will have eight times the shear sensitivity.

A strain gauge mounted on the surface of the shaft will measure linear strain in the direction of the gauge axis. To maximize this for a given shear strain in the shaft, the gauge should be mounted with its active axis at an angle of 45° to the axis of the shaft, as shown in Figure 9. An area of the surface, originally square and with sides of unit length, is deformed by the strain to a parallelogram. The original length of the diagonal is $\sqrt{2}$. If the angle φ is small, the length of the diagonal of the parallelogram will be longer than the diagonal of the square by $\varphi\sqrt{2}$. The longitudinal strain e

in the gauge is thus

$$e = \frac{\varphi/\sqrt{2}}{\sqrt{2}} = \frac{\varphi}{2}.$$

The gauge sensitivity can now be expressed in terms of the strain in the gauge per unit applied torque.

$$\frac{e}{T} = \frac{\varphi}{2T} = \frac{1}{\pi G a^3}.$$

In Figure 9 the gauge suffers tensile strain, but the gauge could equally well be mounted on the other diagonal and would then suffer an equal amount of compressive strain. It is common practice to use both gauges,

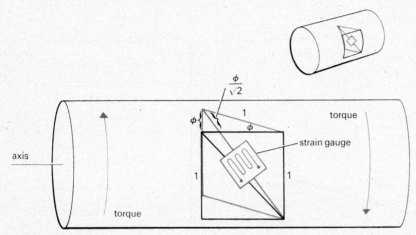

Figure 9   A strain gauge mounted on a torque-sensing shaft at 45° experiences the maximum strain

mutually at right angles and both at 45° to the shaft axis. They are used in the bridge circuit of Figure 10. Overall sensitivity is doubled and temperature compensation is achieved with this arrangement.

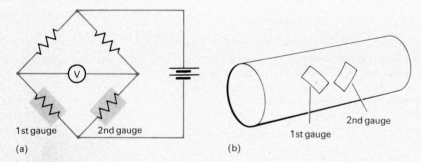

Figure 10(a)   A bridge circuit for measuring torque; (b) the location of the gauges on the shaft

Figure 11 shows a few types of *torque-sensing shaft*. Some are designed so that the angular displacement due to the twisting of the shaft is measured

torque-sensing shaft

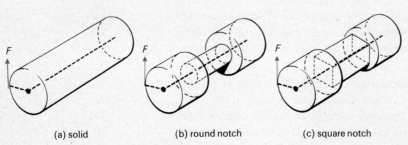

(a) solid   (b) round notch   (c) square notch

Figure 11   Shafts for sensing torque

9

with a displacement transducer. In others, the strain in the surface of the shaft is measured with strain gauges. In the notched shafts, the reduced cross-section in the notched section has a larger strain, giving the transducer increased sensitivity.

If one end of the sensing shaft is rigidly connected to a stationary test frame, so that the whole assembly cannot rotate, the torque measured is called *reaction torque*.

If a transducer is designed to measure torque in a rotating shaft, provision must be made to couple the signal from the secondary transducer mounted on the rotating sensing shaft. This is sometimes done using rotating contacts, called slip rings and brushes, but may also be done using other techniques such as transformer coupling.

reaction torque

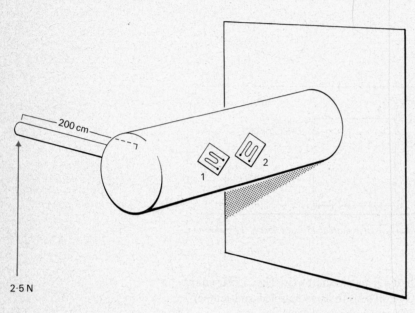

2·5 N

Figure 12   Diagram for SAQ 7

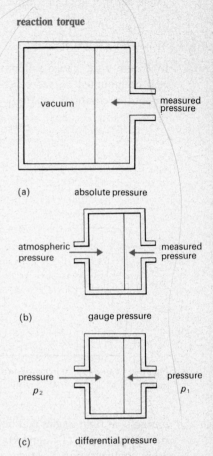

(a)   absolute pressure

(b)   gauge pressure

(c)   differential pressure

Figure 13   The type of pressure measured by a transducer depends upon the pressures on each side of the transducer

SAQ 7

The shaft shown in Figure 12 is rigidly mounted at one end and is subjected to 2.5 N force at the end of a 200 cm rod. It has a shear sensitivity of 0.01 rad $(\text{N m})^{-1}$. Two strain gauges are mounted on the shaft. What is the strain on them, and which one is in tension and which is in compression?

SAQ 7

1.5   Sensing elements for pressure transducers

The pressure on a surface is the force per unit area on that surface. To sense the pressure of a gas or a liquid in a container or pipe, the transducer is attached to that container, so that the liquid or gas presses on the sensing part of the transducer, usually a membrane, plate or cavity, as I shall explain in a moment. The sensing element distorts due to the applied pressure, but the amount of distortion also depends upon the pressure on the other side of the sensing element. As shown in Figure 13 the pressure on the other side determines whether the transducer is measuring gauge, absolute or differential pressure.

The SI unit for pressure is the pascal (Pa). However, at the time of writing (1973), very few manufacturers of pressure-measuring equipment have begun using pascals to calibrate their instruments. In the Appendix I have listed some other units in common use and the conversion factors between them. Only two, the dyne per centimetre squared and the bar, are decimal

10

multiples of the pascal. The others are based upon different systems of units for force or length or upon particular methods of measuring pressure. The units using conventional inches of water (inH$_2$O) and conventional millimetres of mercury (mmHg) are based on the concept of head (Unit 1) using the U-tube manometer (Figure 14), in which pressure is measured by balancing it against the weight of a column of water or mercury.

Pressure transducers are made in a particularly wide variety of forms because of the great number of different applications for them. Apart from the more obvious criteria for selecting a pressure transducer, such as pressure range, accuracy and dynamic response, many applications also require special attention to material, sealing and temperature effects.

The fluid whose pressure is being measured may be corrosive or may be at a high temperature and so introduce large thermal gradients across the transducer.

A number of sensing elements are shown schematically in Figure 15. You have met most of them before, but I have put them in for the sake of completeness. There are diaphragms, either flat or corrugated (to increase the deflection range). The capsules and capsule stacks (not shown) are formed by welding, brazing or soldering two or more diaphragms together to obtain larger deflections. (A diaphragm capsule is used in the pneumatic pressure transducer of Unit 1.) The bellows appear similar to the capsule stacks but are made from a single piece of metal. (Again, used in the pneumatic transducer.)

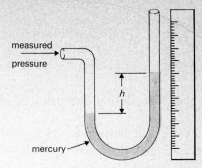

Figure 14   In this U-type manometer the measured pressure (head) is given directly in mmHg by the distance h

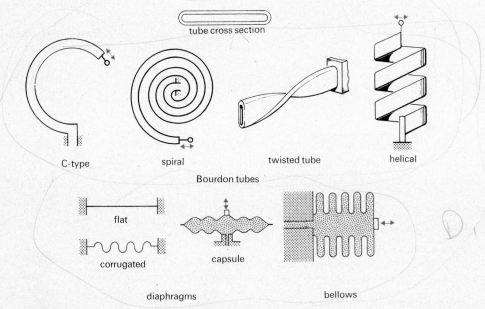

Figure 15   Various elastic pressure-sensing elements

The C-type Bourdon tube is described in Unit 1. The spiral Bourdon is a version in which, because of the increased arc length, increased sensitivity is possible. The twisted tube is inherently more rugged and has similar angular deflection for relatively greater pressures. The helical Bourdon tube, like the spiral, has greater sensitivity than the C-type, in a shape which is more convenient for some applications.

11

Resistive potential dividers (pots)

The resistive potential divider, or 'pot', is the first type of displacement transducer which I shall discuss. Its principle of operation was described in Unit 1, in connection with a rotary type. This section considers the linear type too, and treats them both in greater detail.

2.1 Linear-displacement pots

I shall start by considering the linear-displacement transducer shown schematically in Figure 16. The displacement of the object under test is sensed by connecting it to the shaft of the transducer. In operation, a source of voltage is connected across the terminals labelled + and −. As the shaft moves from its extreme left to its extreme right position the voltage on the wiper varies from that at the − terminal to that at the + terminal. Thus a meter monitoring the wiper voltage can be calibrated in terms of shaft displacement. This is a linear-displacement version of the pot for measuring angular position of Unit 1. The electrical principle is the same.

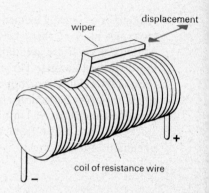

Figure 16   Schematic representation of a pot-type displacement transducer

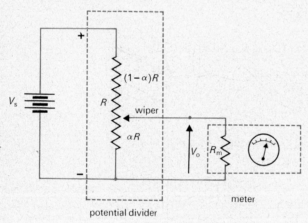

Figure 17   A measuring circuit for a potential divider (pot)

To determine the relationship between shaft position and output voltage, let us consider the circuit of Figure 17. A battery of voltage V_s supplies the pot with current. The meter registers the voltage V_o. The internal resistance of the meter is R_m. The pot has a resistance of R, and that fraction of R below the wiper is aR. The value of a can lie between 0 and 1, and, if the pot is wound uniformly with resistance wire, a varies linearly with the position of the wiper.

For the moment let us assume that R_m is so much larger than R that the current it draws is negligible compared to the total current through the resistive element. To this approximation, the current in the upper and lower parts of the resistance element will be the same, and so the voltage across each part will be proportional to the resistance of that part. Thus the meter voltage V_o will be

$$V_o = aV_s. \tag{1}$$

The meter voltage varies linearly with the position of the wiper, and hence, with the displacement of the shaft.

Because of the current drawn by the meter, the measured voltage is not strictly proportional to the electrical fraction a. The effective resistance R_{eff} below the wiper is the parallel combination of aR and R_m.

What is that effective resistance?

The effective resistance of two resistors in parallel is their product divided by their sum.

In this case it is

$$R_{eff} = \frac{aRR_m}{aR + R_m}. \tag{2}$$

The meter voltage V_m is proportional to the fraction of the total resistance represented by R_{eff},

$$V_m = \frac{R_{eff}}{R_{eff} + (1-a)R} V_s.$$

If I substitute for R_{eff}, using equation (2), this expression becomes

$$V_m = \frac{aV_s}{1 + a(1-a)(R/R_m)}. \tag{3}$$

Equation (3) reduces to equation (1) if $R_m \gg R$. Figure 18 shows equation (3) plotted for various values of R/R_m. If $R_m = 10R$, the maximum error in using equation (1) is about 2.5 per cent and, if $R_m > 10R$, the maximum error is about $25R/R_m$ per cent.

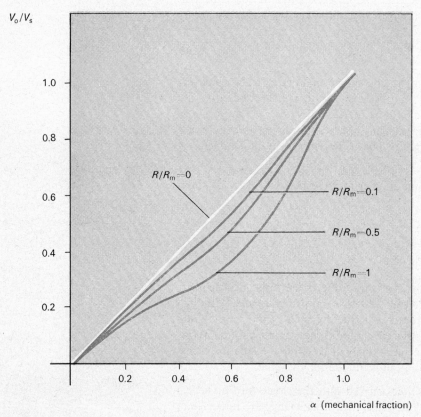

V_o/V_s

$R/R_m = 0$

$R/R_m = 0.1$

$R/R_m = 0.5$

$R/R_m = 1$

α (mechanical fraction)

Figure 18   The output voltage of a potential divider, showing the effect of different relative values of meter resistance

The sensitivity of the transducer is simply found by dividing V_s by the total travel of the transducer. Increasing V_s thus increases the sensitivity, but it also increases the power dissipated by the pot. As all pots have a specified maximum power dissipation, the sensitivity is limited. The power dissipated is given by

$$P = \frac{V_s^2}{R},$$

so, if P_{max} is the rated maximum power dissipation, the maximum applied voltage is

$$V_{s(max)} = \sqrt{(P_{max}R)}.$$

In the following self-assessment question you will see how the various factors just described are inter-related.

SAQ 8

SAQ 8

The following is an extract from a manufacturer's specification for a range of linear pots. Suppose you wish to use one of these pots with a meter with an internal resistance of 10 kΩ and you want a linearity of at least 1 per cent. What is the maximum sensitivity you could obtain? What resistance value would you choose and what voltage would you use to give a maximum sensitivity? (Assume the pots themselves are perfectly linear.)

Specification
Resistance values: 47 Ω, 100 Ω, 220 Ω, 470 Ω, 1 kΩ, 2.2 kΩ, 4.7 kΩ, 10 kΩ, 22 kΩ.
Mechanical travel: 139 mm (0.55 in).
Power rating: 1.5 W at temperatures up to 40 °C.

What, do you think, limits the resolution of this transducer?

The spacing of the turns of the wire, just as in the case of the rotary pot of Unit 1.

In a displacement pot such as the one shown in Figure 16, in which the resistive element is a length of resistance wire coiled around a former (called a *mandrel*), the wiper voltage increases by discrete steps as the wiper contacts successive turns. Thus the resolution is limited to the number of turns per unit length, which is determined by the diameter of the wire used.

Pots are made with a maximum number of turns per centimetre of about 400.

2.2 Other displacement pots

In addition to wire-wound pots, some transducers use resistance elements made of carbon film, conductive plastic or a ceramic–metal mix. Although the resolution of these transducers is limited by such factors as the grain size of the surface of the resistive element and the mechanical limitations of the bearings and wiper, they are often called 'infinite' resolution transducers, implying that the resolution is infinitesimally small.

The resistive element is sometimes curved into the single-turn or helical forms shown in Figure 19. The shaft then senses angular, rather than linear, displacement. Alternatively, a linear-displacement transducer can be made from a rotary pot fitted with a pulley and a cable which is unreeled from it. A spring is used to rewind the cable. This is the arrangement described in Unit 1 for measuring petrol level.

In the first displacement pot we looked at, the fraction a of total displacement was designed to correspond as closely as possible to the fraction of

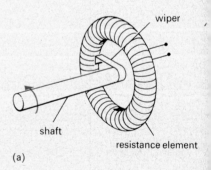

(a)

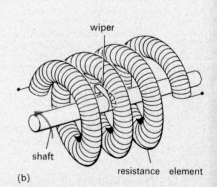

(b)

Figure 19 Schematic diagrams of a circular and a helical potential divider

14

the resistive range of the pot. With some pots using a resistive element made out of conductive plastic or some other continuous material, the fractional displacement of the sensing shaft is made to differ from the electrical fraction by suitably shaping the resistive element. In this way the pot can deliberately be made non-linear. This is usually done to compensate for some other non-linearity in a different part of the measuring system. That is, a non-linear potential divider is used to produce a linear output signal from a non-linear system.

2.3   Transducer example 2: A pressure transducer using a potential divider

A wide variety of pressure transducers are available commercially using displacement pots in combinations with different sensing elements. In this example we shall look at one using a Bourdon tube. Its construction is shown schematically in Figure 21. The operation and internal construction of this transducer are shown in more detail in the television programme 'Pressure transducers'.

Does this transducer measure absolute, gauge or differential pressure?

It measures gauge pressure. The pressure inside the Bourdon tube is measured against atmospheric pressure outside the Bourdon tube.

To reduce the effects of shock and sustained vibration, the inside of the case is filled with silicone grease.

The specification for this transducer is given in Table 2. The power rating is shown in Figure 20. It specifies the maximum power when the winding is uniformly loaded.

When might the winding *not* be uniformly loaded?

If the current drawn through the wiper by the measuring circuit is not a small fraction of the winding current.

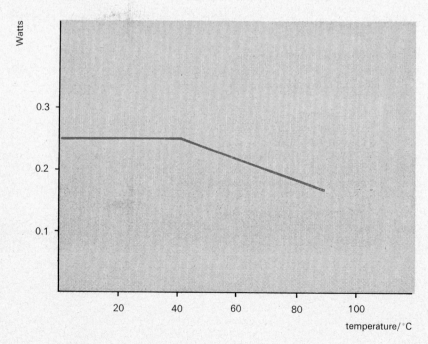

Figure 20   Power rating versus temperature

15

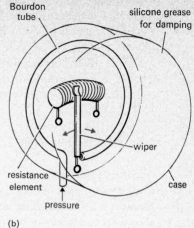

(a)   (b)

Figure 21(a) A Pye Dynamic type PP9 pot-type pressure transducer. (b) A schematic representation of its operation

SPECIFICATION

TABLE 2

Ranges
0–345 kPa (0–50 lb/in²) ;
0–689 kPa (0–100 lb/in²) ;
0–1.37 kPa (0–200 lb/in²) ;
0–3.45 MPa (0–500 lb/in²) ;
0–6.89 MPa (0–1000 lb/in²) ;
0–13.7 MPa (0–2000 lb/in²) ;
0–27.4 MPa (0–4000 lb/in²).

Input resistance at terminals
5 kΩ (nominal).

Power rating
See Figure 20.

Maximum wiper current
10 mA.

End-point settings
Zero pressure end : 0.2–3.0% of the total winding resistance
Full range end : 97.0–99.8% of the total winding resistance.

Insulation resistance
100 MΩ (minimum) at 100 V d.c.

Accuracy (non-linearity and hysteresis)
±1%, except on the 27.4 MPa range, where it is ±1.5%.

Resolution
0.3%.

Working temperature range
From −30 °C to 90 °C.

Thermal effect on span
0.03% per °C.

Life expectancy
10^6 cycles.

Gauge pressure media
Any liquid or gas compatible with beryllium–copper, brass, copper–nickel and silver brazing alloy.

Over-pressure (without calibration shift)
1.2 times rated pressure.

Proof pressure of Bourdon tube
3 times rated pressure.

Weight
0.25 kg.

Section 3

Magnetic transducers

3.1 Introduction

In this section I shall describe another class of displacement transducer which is used both for displacement itself and, in conjunction with sensing elements, for force, torque and pressure measurements. In magnetic transducers, the displacement varies the magnetic properties of the transducer and a corresponding electrical output signal is obtained. The three types included here are:

1 the differential transformer;
2 the variable-inductance transducer;
3 the inductance-bridge transducer.

3.2 The differential transformer

I mentioned a differential transformer in passing in connection with its use with a proof-ring force transducer. Figure 22 shows such a transducer, consisting of a proof ring with a differential transformer, which measures deflection resulting from an applied force. When a load is applied to the top and bottom of the proof ring, their separation changes and the plunger is moved along the main axis of the differential transformer.

This type of differential transformer is sometimes called a *linear variable differential transformer* (*LVDT*) to distinguish it from a type which measures *rotary* displacement, called the RVDT. (Note that the use of the word linear here has nothing to do with the linearity of the relationship between output signal and measurand. It simply refers to the fact that displacement in a straight line is measured.)

The differential transformer of Figure 22 has a primary winding and two secondary windings. The primary is usually fed with a.c. current of constant amplitude and frequency. A.C. voltages are induced in the secondary windings by transformer action.

> You may not be familiar with transformer action and some of the terms such as 'magnetic circuit' and 'coupling'. It is not essential to understand *how* a transformer works to grasp the essentials of the LVDT. All I ask is that you accept that a.c. current in a coil induces an a.c. voltage in an adjacent coil, that is, the two coils are electromagnetically 'coupled'. Ferromagnetic materials (such as iron) introduced between the coils can increase the coupling, thereby increasing the voltage induced in the second coil.

When no load is applied to the proof ring, the plunger is commonly adjusted so that the ferromagnetic armature which it carries is centred in the windings. Because of the resultant symmetrical arrangements, the secondary voltages are then equal.

The series-opposition connection shown in the figure gives an output which is the difference between these voltages. More specifically, $v_o = v_1 - v_2$.* This is zero with the armature centred, because then $v_1 = v_2$.

*The convention I am using is that capital letters are used for d.c. voltages and currents and for amplitudes of a.c. voltages and currents. Lower-case letters are used for time-varying voltages and currents. Thus a sinusoidal a.c. voltage may be written as $v = V_s \sin \omega t$.

linear variable differential transformer
LVDT

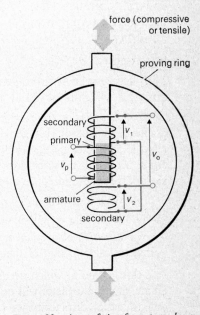

Figure 22 *A proof-ring force transducer using a differential transformer*

When a tensile load is applied to the proof ring, the armature is displaced from its central position and towards the upper secondary winding. Because the armature is made of a ferromagnetic material, the coupling between the primary and the upper secondary coil is increased while the coupling with the lower secondary coil is decreased. Consequently, the voltage induced in the upper secondary is greater in magnitude than the voltage induced in the lower secondary. The difference between the secondary voltages is now no longer zero and an output voltage appears which, over a certain range, is linearly related to the displacement.

Figure 23 shows the waveforms involved. One cycle of the primary voltage is shown in Figure 23(a). With a relatively small tensile load (Figure 23b), v_1 exceeds v_2 slightly in magnitude and, as I have just explained, an output voltage appears, equal to their difference.

Figure 23(c) shows the situation with a larger tensile load. Now v_1 is much greater than v_2 in magnitude, so their difference v_o is larger than in Figure 23(b). Try the following self-assessment question before reading on.

SAQ 9

SAQ 9

Suppose a compressive load is applied to the proof ring so that the armature is deflected towards the lower secondary coil. Sketch v_1, v_2 and v_o for small and large compressive loads.

How can you differentiate between v_o resulting from tension and from an equal compression?

v_o is in phase with the primary voltage for tensile loads and 180° out of phase with the primary for compressive loads.

The way in which the coupling between the primary and the secondaries varies with the armature displacement is not, inherently, linear. But, by careful design of the transformer, an essentially linear variation of the coupling and of the output voltage amplitude can be obtained over the designed operating range. The details of the design for good linearity need not concern us here.

At this point, I shall summarize the essential points of the differential transformer:

1 It has a primary winding, two secondary windings, and a movable armature, which is moved by the displacement to be measured.

2 The secondaries are connected so that, with the armature in a central position, no output voltage is obtained. This 'null balance' arrangement has the same advantage as the connection of strain gauges in a bridge circuit: it avoids the difficulties associated with the measurement of a small variation superimposed on a relatively large constant value.

3 In a well-designed differential transformer the output voltage varies essentially linearly with displacement over the operating range.

4 The *direction* of displacement is evident from the phase of the output voltage relative to that of the primary voltage.

It is this last point which needs elaboration. If the output signal is fed to an a.c. meter, the meter will not be able to distinguish between two signals due to equal displacements in opposite directions, because the meter responds only to the amplitude of the signal. It has no way of determining that one signal happens to be out of phase with the primary voltage.

In order to distinguish between two output signals which have the same amplitude but opposite phase, some form of signal conditioning must be used. One common method is to use a circuit called a phase-sensitive detector, which is described in the next section.

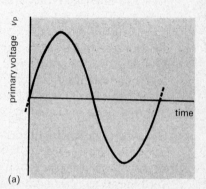

(a)

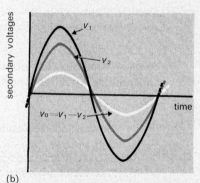

(b)

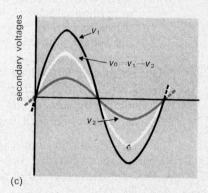

(c)

Figure 23   Differential transformer voltage waveforms showing secondary voltages for (b) a small tensile force and (c) a large tensile force

Performance limitations

The resolution of a differential transformer in the region where the arma-ture is centred between the two secondary windings depends upon how accurately the voltages in the secondaries balance one another. In practice they do not cancel completely and so a non-zero minimum, or *null voltage*, arises. One of the reasons is that slight asymmetries in the construction of the windings, bobbin and core cause a slight phase difference between the voltages induced in the two secondary windings. Thus, although their amplitudes may be equal, the value of their difference is not always zero.

null voltage

3.3 The phase-sensitive detector

When used with a differential transformer, a phase-sensitive detector enables one to distinguish between output voltages corresponding to displacements of the armature on either side of its centre position.

Let us be clear about what the inputs to a phase-sensitive detector are:

1 The sinusoidal output voltage of the differential transformer. As shown in Figure 24 (for the force transducer just discussed), the output voltage is either in phase or out of phase with the primary voltage of the differential transformer.

2 The primary voltage (in certain types only).

The output of the phase-sensitive detector should be an electrical signal which is proportional in magnitude to the amplitude of the differential transformer output voltage v_o, and which is positive when v_o is in phase, and is negative when v_o is out of phase with the primary voltage.

I would like to describe the operation of one of the many circuits used as phase-sensitive detectors.

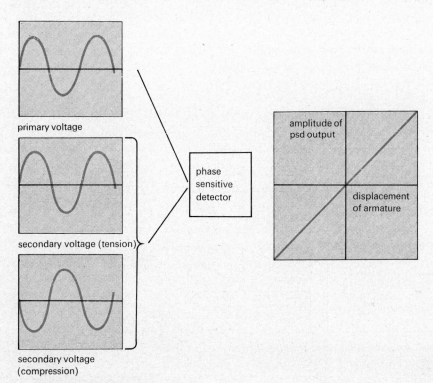

Figure 24 The inputs to a phase-sensitive detector and the curve of its output versus displacement

Figure 25 shows a simplified diagram of this circuit. The phase-sensitive detector consists of an amplifier and an electronic switch. The switch position is controlled by the primary voltage of the differential transformer.

The switch is in position 1 when the primary voltage is positive and in position 2 when it is negative. Thus v_d, the detector output voltage, is equal to the output voltage v_o of the differential transformer when the primary voltage is positive, and is equal to $-v_o$ when the primary voltage is negative.

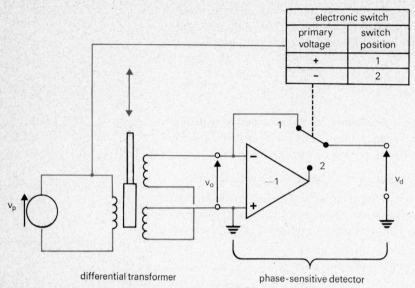

electronic switch	
primary voltage	switch position
+	1
−	2

differential transformer   phase-sensitive detector

Figure 25   A simplified diagram of one type of phase-sensitive detector

The resulting output waveforms v_d are shown in Figure 26. When v_o is in phase, its negative half-cycles occur when the switch is in position 2 and so are inverted by the amplifier. The resulting v_d thus consists entirely of positive half-sinusoids only. Its average value is positive and is proportional to the amplitude of v_o.

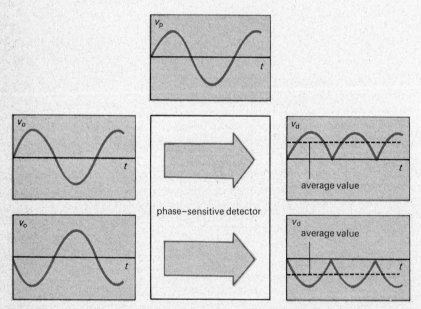

Figure 26   The inputs and corresponding outputs of the phase-sensitive detector

Similarly, if v_o is out of phase with v_p, the corresponding v_d consists of negative half-sinusoids only. Its average value is again proportional to the amplitude of v_o, but its sign is now negative.

In a practical realization of this type of phase-sensitive detector the switch is at the input to the amplifier instead of the output. Instead of switching the input signal either *through* or *around* the amplifier (as in Figure

25), the input signal is always fed into the amplifier, but the switch changes the feedback network so the gain of the amplifier is changed either to +1 or −1.

SAQ 10

SAQ 10

With the armature in its centre position, the null voltage appearing at the output of a differential transformer may appear as shown in Figure 27. Notice that it is 90° out of phase with the primary voltage. Sketch the detector output which would appear if this null voltage is fed into a phase-sensitive detector. What is its average value?

3.4 Transducer example 3: A differential transformer for measuring displacement

Figure 30 is a photograph of a displacement transducer whose specification is given in Table 3. Both its input and its output voltages are d.c., so it can be used analogously to a pot.

To do this there are some electronic circuits built into the case along with the differential transformer. This is shown schematically in Figure 28. The oscillator is a circuit which generates a.c. (In this case it generates a square wave rather than a sinusoid.) It also contains a phase-sensitive

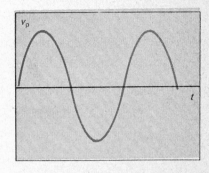

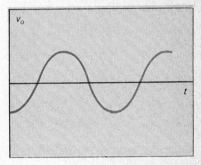

Figure 27   Diagram for SAQ 10

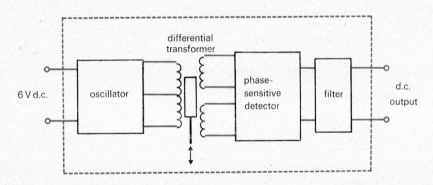

Figure 28   Block schematic of the transducer

detector and a *filter* circuit. The effect of the filter is to 'smooth' the output of the phase-sensitive detector. At any steady displacement in the linear region, the filter output is a nearly constant voltage proportional to the average value of the filter input. The remaining varying part of the output voltage is called the *ripple voltage* (see Figure 29) and is less than 1 per cent of the output voltage.

As the armature is moved, the average value of the output voltage varies. To ensure that the filter does not 'smooth' these variations (as they are what is being measured) the rate at which the armature moves must be much slower than the oscillator frequency. The oscillator frequency is 7–10 kHz. The armature movement should be within the frequency range of the 'filtered output bandwidth' which is 0–75 Hz.

The transducer requires a d.c. input voltage between 6 V and 12 V. The full-scale output voltage obtained is proportional to the supply voltage. For a 6 V input the full-scale output voltage is ±2 V, giving a total output voltage range of 4 V. Thus, it is not necessary to use a very sensitive voltmeter on the output. Since the sensitivity is just the full-scale output voltage divided by the full-scale displacement, it too varies in proportion to the input voltage.

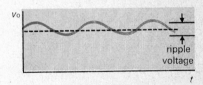

Figure 29   Filter output voltage. The ripple has been exaggerated for clarity

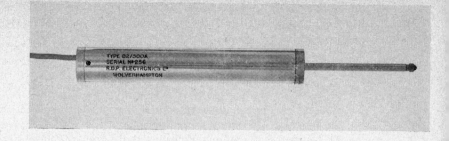

SPECIFICATION

TABLE 3

Supply voltage
6 V d.c. (50 mA), 12 V max.

Output sensitivity
1.6 V cm^{-1}.

Displacement range
12 cm (0.5 in) either side of its centre position.

Internal oscillator frequency
7–10 kHz.

Linearity
Better than 0.5% of working range.

Temperature coefficient of sensitivity
0.5% full-scale output/°C (typical).

Zero shift with temperature
0.05% of full-scale output/°C (typical).

Noise output at zero displacement
50 mV (typical).

Filtered output bandwidth
0–75 Hz (at 75 Hz the output will be reduced by the filter by a factor of 0.7).

Operating temperature range
From −10 °C to 50 °C (85 °C absolute maximum continuous temperature).

Length
132 mm.

Total weight
170 g.

Figure 30   An RDP Electronics type D2/500A d.c. differential-transformer displacement transducer

3.5 Variable-inductance (variable-reluctance) displacement transducers

An example of an inductive displacement transducer is shown schematically in Figure 31.

The central limb of the E-shaped piece carries a coil carrying a.c. current. This coil sets up an alternating magnetic flux which follows the path, shown in the figure, through the ferromagnetic E-piece, the air gaps and the ferromagnetic plate, which is moved relative to the E-piece by the displacement to be measured. (The ferromagnetic material is commonly steel.) When the plate is moved, it changes the properties of the magnetic flux path. As a result, the current in the coil changes. This current forms the output signal.

The transducer works as follows. The voltage v in a coil carrying a current i is proportional to the rate of change of that current. That is,

$$v \propto \frac{di}{dt}$$

or $\quad v = L\dfrac{di}{dt}.$

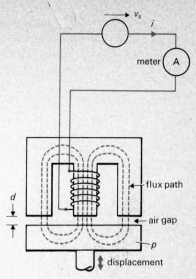

Figure 31   Schematic representation of an inductive displacement transducer

The proportionality constant L is called the *inductance* of the coil. The unit of inductance is the *henry* H. If the current is sinusoidal, of the form

$$i = I \sin \omega t$$

the voltage will be

$$v = L(\omega I \cos \omega t).$$

inductance
henry

The amplitude V of the voltage is $L\omega I$, or $L\omega$ times the amplitude of the current. Alternatively, the amplitude I of the current is $V/L\omega$. Thus, if the amplitude of the voltage across the coil is fixed, the amplitude of the current varies inversely as the inductance.

Moving the plate of the transducer in Figure 31 varies the inductance of the coil, hence the name 'variable-inductance transducer'. The inductance of the coil depends upon the properties of the magnetic circuit formed by the E-shaped piece on which the coil is wound, the movable plate and the air gaps. A detailed discussion of magnetic circuits is beyond the scope of this course. However, an introduction to them is given in the radio programme 'Magnetic circuits'. For the purposes of this section it is sufficient to know that the inductance of a coil with N turns of wire is related to the *reluctance* R_m of the magnetic circuit by

reluctance

$$L = \frac{N^2}{R_\mathrm{m}}.$$

The reluctance is that property of the magnetic circuit which, for a given current in a given coil, determines how much magnetic flux is produced. The reluctance determines the flux in a magnetic circuit analogously to the way resistance in an electrical circuit determines the current. The greater the reluctance, the smaller the flux.

In the transducer in Figure 31 the reluctance of the plate and of the E-shaped piece remains substantially constant as the plate is moved. However, the reluctance of the air gap is proportional to its width d, at least over a limited range. Thus, changing d changes the reluctance. That in turn changes the inductance, and so, for a voltage of a constant amplitude, the amplitude of the current changes. The current meter can thus be calibrated in terms of d.

The transducer of Figure 31 has its own plate moved by the displacement to be measured, and is calibrated appropriately.

On the other hand, the transducer of Figure 32 is intended for use in proximity to a metallic surface. This might be ferromagnetic (e.g. steel)

23

and decrease the transducer's reluctance, as with the transducer of Figure 31, or it might be such as to *increase* the reluctance (brass or aluminium, for example). Since the magnetic properties of the surface in proximity are

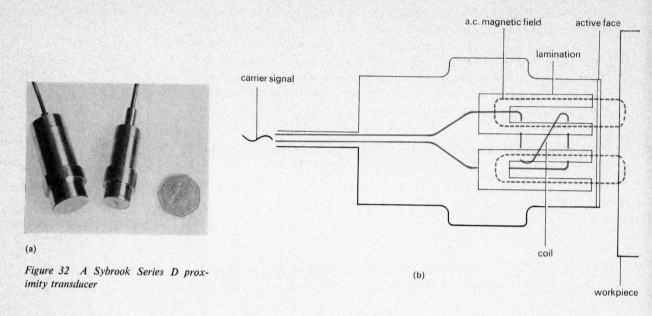

(a)

*Figure 32   A Sybrook Series D prox-
imity transducer*

(b)

unpredictable, it is impossible to calibrate this transducer for all possible cases, and it serves more as a proximity detector than as a transducer in our sense.

SAQ 11

SAQ 11

The electrical source for the variable inductance transducer in Figure 33 supplies an a.c. current with a constant amplitude. The displacement of the plate is determined by measuring and appropriately calibrating the amplitude of the voltage v_L across the transducer. How does the sensitivity of this transducer (the voltage change per unit change in displacement) vary with the frequency of the current? Does the voltage increase or decrease with increasing displacement?

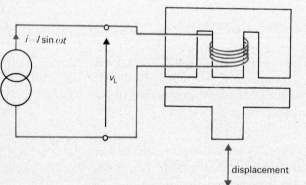

*Figure 33   An inductive displacement transducer driven by a constant-amplitude
current source has v_L as its output*

3.6   Transducer example 4: An inductance-bridge pressure transducer

Figure 34 shows a pressure transducer in which the two input pressures act upon either side of a diaphragm. The centre of the diaphragm is displaced, to either left or right, in proportion to the differential pressure.

24

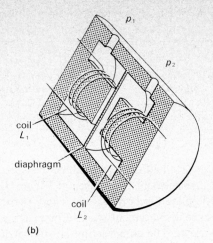

(a)   (b)

SPECIFICATION

TABLE 4

Operating range
19 ranges available from 0–3.45 kPa (0–5 p.s.i.) to 0–3.45 MPa (0–5000 p.s.i.). All information below applies to the 0–3.45 MPa model.

Maximum case pressure
82.6 MPa (12 000 p.s.i.).

Temperature range
From 0 °C to 100 °C.

Temperature error of span
0.03%/°C.

Resonant frequency of diaphragm
4 kHz.

Mechanical shock
1000 g for 1 ms in each of three mutually perpendicular axes will not affect calibration.

Terminal non-linearity
Less than 0.4% of full range.

Side-to-side error
(Change of slope at zero differential pressure) : 1% of full range

Hysteresis
Less than 0.4% of full range.

Zero shift with pressure reversal
1% of full range.

Resolution
Infinite (i.e. limited, in fact, by the resolution of the electronics used with the transducer).

Output
25 mV ±50% per volt of excitation.

Maximum supply voltage
10 V r.m.s., 1–10 kHz (3 kHz optimum).

Coil resistance
75 Ω approximately.

Coil inductance
15–20 mH.

Insulation (coils to case)
100 MΩ at 250 V d.c.

Internal volume
0.33 cm³.

Weight
84 g.

Figure 34   An SE Laboratories Type SE74 inductance-bridge pressure transducer

What is the differential pressure?

The differential pressure is the *difference* between the two input pressures.

If $p_1 > p_2$, which way will the diaphragm be displaced from its position when $p_1 = p_2$?

The diaphragm will be displaced to the right in Figure 34.

To either side of the diaphragm, each half of the transducer has essentially the same construction as the displacement transducer of Figure 31. The diaphragm is the ferromagnetic plate for both halves. These two halves, then, work as variable-inductance displacement transducers, each sensing the displacement of the diaphragm. The construction is symmetrical, so the inductances L_1 and L_2 of the two coils are equal when the diaphragm experiences zero differential pressure.

The two coils are connected into a bridge circuit, as in Figure 35. The principle of the bridge circuit is just the same as that of the strain gauge bridges of Unit 2. That is, each resistor drops half the a.c. bridge energizing voltage and the two inductors, being equal for zero differential pressure, do the same. So the voltages either side of the meter are equal, and the meter reading is zero, for zero differential pressure input. Again, a symmetrical transducer in conjunction with a bridge circuit results in zero output for zero measurand. If p_1 exceeds p_2, the diaphragm is displaced towards the right, decreasing the reluctance (because of the reduced air gap) and hence increasing the inductance ($L = N^2/R_\mathrm{m}$) of L_2. At the same time L_1 decreases.

Now, I showed previously that, in a coil, the amplitude I of the current is equal to the amplitude V of the voltage divided by ωL. That is,

$$I = \frac{V}{\omega L}.$$

Both sides of the transducer carry the same current, so an increase in L_2 and a decrease in L_1 are accompanied by an increase in the voltage across L_2 and an equal decrease in the voltage across L_1. (The sum of the two voltages equals the bridge input voltage, which has fixed amplitude.)

As a result, the voltage across the meter in Figure 35 increases from zero. This voltage is, of course, a.c. Figure 36 shows the sinusoidal waveforms involved. The bridge energizing, or input, voltage is shown at (a). The situation with $p_1 > p_2$, just described, causes the output voltage shown at (b). Note that when the bridge input voltage is instantaneously positive, so is the voltage at (b), as you should expect when the voltage across L_2 exceeds that across L_1. When p_1 falls *below* p_2, the situation is reversed. L_2 becomes less than L_1 and, for an instantaneously positive bridge input voltage, the output voltage becomes, instantaneously, *negative*. The waveform at (c) represents this output, which is an inverted version of that at (b).

If the transducer is used with differential pressure of only one sign, that is, if p_1 *always* exceeds p_2, or p_2 *always* exceeds p_1, this change of phase of the output waveform cannot occur, and an a.c. voltmeter will suffice to indicate the differential pressure.

If, however, the application allows the input differential pressure to change sign, then a phase-sensitive detector, such as that used with the differential transformer, is necessary in order to distinguish between positive and negative differential pressures.

The internal construction of this pressure transducer is shown in the television programme 'Pressure transducers'.

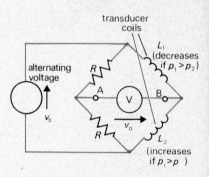

Figure 35 *An a.c. bridge circuit can be used to detect the relative changes in the inductances L_1 and L_2*

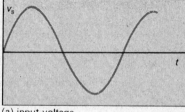

(a) input voltage

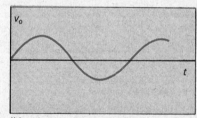

(b) v_o for $p_1 > p_2$

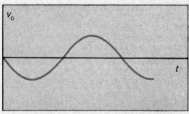

(c) v_o for $p_1 < p_2$

Figure 36 *Voltage waveforms for the inductance bridge circuit of Figure 35: (a) the input voltage; (b) the output voltage v_o for $p_1 > p_2$; (c) the output voltage v_o for $p_1 < p_2$. Note the difference in phase between (b) and (c)*

26

Section 4

Capacitive transducers

4.1 Capacitors and capacitance

In Sections 2 and 3 I discussed secondary transducers which convert mechanical displacements into electrical signals, using the potential-divider principle and magnetic effects respectively. One important difference between these two types is that some of the latter type did not require any mechanical contact between the moving and stationary parts of the transducer. In this section we shall look at another category of secondary transducers, *capacitive transducers*, which share this non-contacting property.

A capacitor consists of two conducting plates (usually metallic) separated by an insulator, which may be air. One form of construction, where the two plates are flat and parallel, is shown in Figure 37.

When a voltage is applied to the metal plates of a capacitor, such as in Figure 37, equal and opposite electric charges appear on the plates. The ratio of that charge to the voltage is the *capacitance*. That is, $q/v=C$, where q is charge (SI unit: coulomb C), v the voltage (SI unit: volt V) and C the capacitance (SI unit: farad F). The capacitance of a parallel-plate capacitor, such as that of Figure 37, is proportional to the area A of the plates and inversely proportional to their separation d. That is,

$$C=\frac{\varepsilon_0 A}{d}.$$

The constant of proportionality ε_0 is the *permittivity* of free space, that is, of a vacuum. Its value is 8.854×10^{-12} F m^{-1} (8.854 pF m^{-1}). Air has a permittivity which may be considered, for our purposes, to be of the same value as ε_0.

capacitance

permittivity

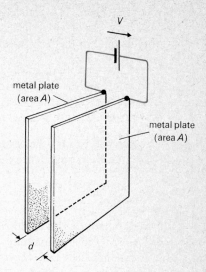

Figure 37 *A parallel-plate capacitor*

4.2 Capacitance variation caused by displacement

Displacements which vary either the effective area of the plates or their separation vary the capacitance. Figure 38(a) shows another

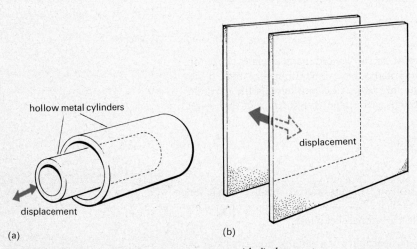
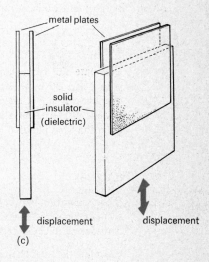

(a) (b) (c)

Figure 38 *Three methods of varying capacitance with displacement*

type of capacitor using concentric hollow metal cylinders. The capacitance of this type, just like the flat plate type, is proportional to the area of overlap of the 'plates'. Because this area of overlap varies linearly with displacement along the axis of the cylinders, the capacitance varies linearly with displacement too.

Figure 38(b) shows a parallel-plate capacitor where the plate separation is varied by the displacement. In this case, the capacitance is inversely proportional to the displacement, because $C = \varepsilon_0 A/d$.

In the first case the variation is essentially a linear function of the displacement, and in the other case the variation is inversely proportional. But whether the overall measurement of displacement is linear depends on how the capacitance is measured, as I shall show shortly.

Another way to vary the capacitance is to introduce an insulator between the plates. The capacitance then becomes

$$C = \frac{\varepsilon_0 \varepsilon_r A}{d}.$$

The extra term in this expression, ε_r, is the permittivity of the insulator, or dielectric as it is commonly known, relative to the permittivity of free space. It is called the *relative permittivity*. To vary the capacitance, an insulator almost as thick as the plate separation is moved into the space between the plates (Figure 38c) changing the proportion of the plate area A affected by ε_r.

relative permittivity

All these methods of converting displacement to variations in capacitance are used in transducers, the particular method in each case being determined by such factors as the amount of displacement involved and the linearity requirements.

For instance, for a given plate area, small values of d, the plate separation, give relatively large values of capacitance. Small displacements about a small d, then, give relatively large *changes* in capacitance, which are relatively easy to measure. So small displacements might best be measured by allowing them to change capacitor plate separation, providing the non-linear displacement–capacitance relationship is not an over-riding disadvantage.

On the other hand, relatively large displacements are best made, generally speaking, with a variable-area capacitor such as that of Figure 38(a). The variable-area type has an essentially linear variation of capacitance with displacement. But the circuit used for measuring its capacitance must produce an output signal linearly related to capacitance if linearity between displacement and output signal is to be achieved for this type.

SAQ 12

SAQ 12

Calculate the capacitance of an air-spaced parallel-plate capacitor with plates 30 mm × 30 mm and a plate separation of 1 mm. Take $\varepsilon_0 = 8.85$ pF m^{-1}. Calculate the *change* in capacitance as the movable plate is displaced, first 0.5 mm nearer, and then 0.5 mm farther away from the fixed plate.

SAQ 13

SAQ 13

If the separation between the two cylinders of a variable-area capacitor is small compared to their radii, its capacitance can be approximated by considering the cylinders as curved parallel plates.

Calculate the total change in capacitance of a variable-area capacitor of the type of Figure 38(a) as the inner cylinder is moved from a position completely overlapping the outer cylinder to a position completely removed from it. The length of each cylinder is 300 mm. The radii of their adjacent surfaces differ by 0.5 mm, and their *mean* radius is 20 mm.

One point which emerges from the calculations in SAQs 12 and 13 is that the capacitance values encountered in capacitive transducers are relatively small compared with the capacitance of typical connecting cables, which may have capacitances in the range 30–100 pF per metre length of cable. This cable capacitance appears across the transducer terminals and adds to the capacitance of the transducer. Clearly, changing the length of cable connecting the transducer to the measuring circuit would change the total capacitance connected across the transducer terminals and thereby change its capacitance–displacement relationship.

For this reason, it is best to place the capacitance-measuring circuit close to the transducer, so that the connecting wires are short and have relatively low capacitance. Where a remote display or recording of displacement is required, the connecting cable should carry the output signal from the capacitance-measuring circuit, rather than from the transducer, and the system will then operate at the same sensitivity regardless of the cable length.

Even with short cable lengths, the capacitance of the cable and 'stray' capacitance between the transducer terminals and other metallic parts of its structure still form a substantial part of the total capacitance of a transducer. This means that, in some capacitive transducers, the change in capacitance resulting from a full-range displacement is a small fraction of the total capacitance.

4.3 Capacitance-measuring circuits

Suppose a capacitive transducer is subjected to a steady displacement and has a constant voltage applied, as in Figure 37. Its capacitance could be measured using the relationship $C = q/v$, provided that both q and v could be measured. Unfortunately, it is impossible to measure the stored charge q without disturbing it in some way. If the voltage v is made to vary, then q must vary too, and its *rate of change* can be measured, for the rate of change of charge is simply current, that is, $dq/dt = i$. A flow of charge of one coulomb per second is called a current of one ampere.

Differentiating $q = Cv$

gives $\qquad \dfrac{dq}{dt} = C\dfrac{dv}{dt}.$

But $\qquad \dfrac{dq}{dt} = i,$

so $\qquad i = C\dfrac{dv}{dt}.$

If the applied voltage is sinusoidal,

$\qquad v = V \sin \omega t,$

so $\qquad \dfrac{dv}{dt} = V\omega \cos \omega t$

and $\qquad i = V\omega C \cos \omega t.$

So the *amplitude I* of the a.c. current through the capacitor is proportional to the capacitance

$\qquad I = V\omega C.$

I said previously that the changes in capacitance resulting from full-range displacement might be a small fraction of the total capacitance. There is a relatively small change superimposed on a relatively large value, rather as in the case of the strain gauge.

That is not an ideal situation for accurate measurements. One solution lies in the use of a symmetrical arrangement, a differential capacitance transducer made up of two capacitors in much the same way as the in-

29

ductance-bridge transducer is made up of two inductances. The differential capacitance transducer is used as part of an a.c. bridge circuit.

Figure 39 shows differential forms of the variable-separation and variable-area types of capacitive transducer. In each case, two capacitors are

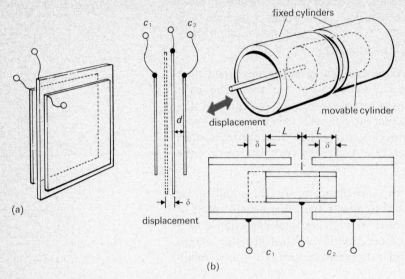

(a)

displacement

(b)

Figure 39   Differential-capacitance displacement transducers

formed, which have equal capacitances when the transducer is geometrically centred. Displacement from this central position increases one capacitance and lowers the other.

The two capacitors of the transducer are connected into a bridge circuit such as that of Figure 40. Difference in capacitance caused by displacement unbalances the bridge and an a.c. output voltage occurs.

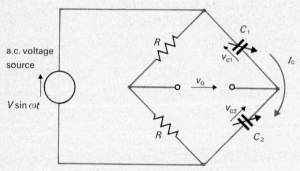

Figure 40   An a.c. bridge for differential-capacitance measurement

The open-circuit bridge output voltage can be calculated as follows. In the bridge circuit the voltage across each capacitor can be found from the relationship $I = V\omega C$, which, applied to the bridge circuit in Figure 40, gives

$$V_{C_1} = \frac{I_C}{\omega C_1} \quad \text{and} \quad V_{C_2} = \frac{I_C}{\omega C_2},$$

because the same current flows through both capacitors. So the voltage across C_2, as a fraction of the bridge energizing voltage V, is

$$\frac{V_{C_2}}{V} = \frac{V_{C_2}}{V_{C_1} + V_{C_2}} = \frac{I_C/\omega C_2}{I_C/\omega C_1 + I_C/\omega C_2}$$

$$= \frac{1/C_2}{1/C_1 + 1/C_2}$$

$$= \frac{C_1}{C_2 + C_1}.$$

30

The variable-separation type

The capacitances of C_1 and C_2 are equal when the central plate is centred. If this balance capacitance is C_0,

$$C_0 = \frac{\varepsilon_0 A}{d},$$

where d is the balance separation in Figure 39(a). When the central plate is displaced by a distance δ,

$$C_1 = \frac{\varepsilon_0 A}{d-\delta} = C_0 \frac{d}{d-\delta}.$$

Similarly, $C_2 = C_0 \dfrac{d}{d+\delta}.$

Substituting C_1 and C_2 into the expression for V_{C_2}/V yields

$$\frac{V_{C_2}}{V} = \frac{d+\delta}{(d-\delta)+(d+\delta)} = \frac{d+\delta}{2d}.$$

The voltage across each resistor in the bridge circuit is half the energizing voltage, so the amplitude of the output voltage is

$$V_o = V_{C_2} - \tfrac{1}{2}V$$
$$= \left[\frac{d+\delta}{2d} - \frac{1}{2}\right]V$$
$$= \frac{\delta V}{2d}.$$

So, although each of the variable-separation capacitors in the transducer has a non-linear capacitance–displacement relationship, the differential arrangement connected into a bridge circuit yields an intrinsically linear relationship between displacement and bridge a.c. open-circuit output voltage.

The variable-area type

In this case, the capacitances C_1 and C_2 (Figure 39b) are proportional to their active areas, so

$$C_1 = \frac{l+\delta}{l}C_0 \quad \text{and} \quad C_2 = \frac{l-\delta}{l}C_0.$$

Substituting C_1 and C_2 into the expression for V_{C_2}/V yields

$$\frac{V_{C_2}}{V} = \frac{l+\delta}{(l-\delta)+(l+\delta)} = \frac{l+\delta}{2l}.$$

The amplitude V_o of the bridge open-circuit output voltage is

$$V_o = V_{C_2} - \tfrac{1}{2}V$$
$$= \left[\frac{l+\delta}{2l} - \frac{1}{2}\right]V$$
$$= \frac{\delta V}{2l}.$$

So the variable-area type also produces a linear output.

Both the results obtained here can be expressed as

$$V_o = \frac{x}{2}V,$$

where x is the fractional displacement (δ/d or δ/l), a result which resembles those quoted in Unit 2 for strain-gauge bridges.

Other bridge arrangements are sometimes used instead of that of Figure 40. Whatever the bridge circuit, the direction of the displacement in a

Figure 41   A Rosemount Engineering Model E1151 DP differential-capacitance pressure transducer

SPECIFICATION

Input
From liquid, gas, or vapour, in three pressure ranges:
From 0–1.25 to 0–7.5 kPa (from 0–125 to 0–750 mm water).
From 0–6.3 to 0–38 kPa (from 0–630 to 0–3800 mm water).
From 0–32 to 0–190 kPa (from 0–3.2 to 0–19 m water).

Elevation and suppression
Zero may be suppressed by up to 50 % of calibrated span, or elevated by up to 150 % o
calibrated span, but span may not exceed ±100 % of maximum range of the particular model

Overpressure limits
From full vacuum to 14 MPa on either side without damage to the transducer. The flanges
will withstand up to 70 MPa proof pressure.

Volume displacement
Less than 0.15 ml.

Initial adjustment accuracy
To ±0.2 % of calibrated span including combined effects of linearity, hysteresis and
repeatability. The linearity will be typically ±0.1 % of calibrated span, repeatability ±0.05 %
of calibrated span and hysteresis 0.05 % of span. (0–1.25 to 0–7.5 kPa models will be
0.1 % of span.)

Stability
Better than ±0.2 % of maximum span for six months.

Ambient temperature effect
(At minimum span. e.g. 0–6.3 kPa for 0–6.3 to 0–38 kPa range.)
The zero error will not exceed ±0.06 % of calibrated span per degC change in ambient.
Total effect including span and zero error will not exceed ±0.07 % of calibrated span per
degC change in ambient.

Temperature limits
Amplifier housing: from −25 °C to 90 °C.
Sensor module: from −40 °C to 105 °C.
Storage temperature: from −50 °C to 120 °C.

Overpressure effect
Overpressure of 14 MPa will shift the zero by less than ±0.25 % of maximum span for low- and medium-range models. For 0–19 kPa range models, the zero shift will be less than ±1.0 % of maximum span limit. The effect on span will be negligible.

Vibration effect
The effect of vibration will not exceed ±0.05 % of maximum span limit per g to 200 Hz in any axis up to a maximum of $2g$.

Mount position effect
A maximum zero shift of up to 250 Pa (25 mm water) may occur due to mounting position. This effect can be calibrated out. There is no effect on the span calibration. Tilt in the plane of the diaphragm only, has no effect on zero or span calibration.

Humidity
The transducer is designed for operation over 0–100 % RH.

Time constant
The transducer has a fixed response time of 0.2 s determined by the hydraulic coupling in the sensor module.

Power supply
For 4–20 mA d.c. output: minimum 14 V d.c.,
maximum 45 V d.c., at the transducer terminals.
For 10–50 mA d.c. output: minimum 30 V d.c.,
maximum 85 V d.c., at the transducer terminals.

Power supply effect
Output will change less than 0.005 % per volt change in supply voltage, measured at the transducer terminals.

differential capacitor determines the phase of the output relative to that of the energizing voltage, just as in the cases of the differential transformer and the inductance-bridge transducer. One way of sensing this phase is to use a phase-sensitive detector, although other methods, outside the scope of this course, are also used. It is common practice for manufacturers of capacitive transducers to supply electronic signal-processing equipment specially designed for their transducer. Linearizing and temperature-compensating circuits are built in, and the output is a d.c. voltage or current analogue signal.

4.4 Transducer example 5: A capacitive differential pressure transducer

Figure 41 shows a photograph and specification of a differential pressure transducer using a variable-separation capacitor of the differential type. The internal construction is shown in Figure 42. The pressures acting on the isolating diaphragms set up similar pressures in the silicone oil filling the space between them. A net force proportional to the difference between the two pressures acts upon the metal sensing diaphragm and deflects it to one side or the other depending on which input pressure is the greater. On either side of the sensing diaphragm are two fixed capacitor plates, held in place by rigid insulation. Each plate forms a capacitor with the sensing diaphragm, which is connected electrically to the metallic body of the transducer.

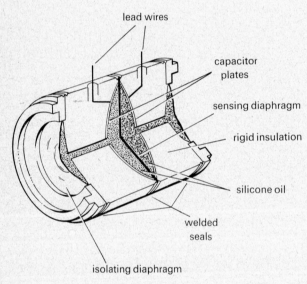

Figure 42   Internal construction of the transducer

The sensing diaphragm and capacitor thus form a differential variable-separation capacitor. When the two input pressures are equal, the diaphragm is positioned centrally and the capacitances are equal. A difference in the two input pressures causes displacement of the sensing diaphragm and is sensed as a difference between the two capacitances.

Electronic circuitry is built into the circular housing above the capacitor housing. It compares the two capacitances and delivers an output signal in the form of a d.c. current, varying linearly from 4 to 20 mA over the operating range. Alternatively, a version delivering from 10 to 50 mA is available. Another model has an evacuated chamber surrounding one of its isolating diaphragms, and measures the *absolute pressure* acting upon the other isolating diaphragm via the single input port.

Yet another model has the chamber surrounding one isolating diaphragm vented to the atmosphere, so it measures the *gauge pressure* at its one input port.

Section 5

Piezoelectric transducers

5.1 Piezoelectricity

The last type of transducer to be dealt with in this unit differs in one rather fundamental aspect from the others: it is the only type which needs no electrical supply. An applied force or displacement causes the transducer to generate on output voltage by the *piezoelectric effect*.

When a quartz crystal is cut so as to have parallel faces, as in Figure 43, and is then squeezed between these faces, electric charges of opposite polarity appear on the faces. The amount of charge depends linearly on the force applied, within a limited range, and its polarity depends on the directions of the crystallographic axes. (For interest only, the quartz is cut so that its crystallographic x-axis is normal to the faces, and it is the direction of this axis which determines the polarity). Certain other crystals, and certain ceramic compounds, such as lead zirconate–titanate, are also used in piezoelectric transducers.

Whereas the crystals, such as quartz, are single crystals with a uniform unidirectional crystallographic structure throughout, the ceramic compounds are polycrystalline and their constituent crystals have random crystallographic orientation. To give them piezoelectric qualities, they are first raised above a temperature called the Curie temperature and then allowed to cool to room temperature while an electric field is applied. The ceramic is then said to be electrically polarized. The application of the field is most easily done by applying a d.c. voltage across metallic electrodes, typical silvered, which are plated onto the two faces.

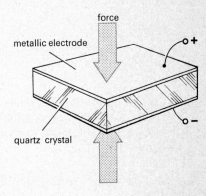

Figure 43 The piezoelectric effect

Metallic electrodes are used with both types, crystal or ceramic, to permit the charge resulting from an applied force to be measured. The crystal or ceramic itself is an insulator. Thus, together with the plates, they form a capacitor.

> What voltage appears across the plates for a charge q if the capacitance is C?
>
> $v = \dfrac{q}{C}$.

5.2 Piezoelectric transducers

Both quartz crystals and ceramics have some of the highest elastic constants, or moduli, of all solids. A relatively large force causes relatively little compression of the piezoelectric material. So piezoelectric transducers are essentially *force* transducers. They can be used as displacement transducers only in cases where the mechanical system which causes the displacement is not significantly affected by the stiffness of the transducer.

Piezoelectric instrumentation transducers have cases in which the crystal or ceramic is mounted, together with metal plates to protect the faces of the piezoelectric material. They include a cable to bring out the output voltage. Sometimes they act as a secondary transducer within a transducer, where a measurand such as acceleration is first transduced to force. In these cases they are built into the main transducer and their construction is integral with it. You will see an example in Units 8/9/10.

5.3 The step response of a piezoelectric transducer

In this section I shall describe the way in which the output voltage responds to a step force input when the transducer is electrically loaded by a voltage-measuring circuit. Figure 44 represents a piezoelectric transducer by its capacitance C, and R represents the resistance of a measuring circuit such as a voltmeter or amplifier input.

Suppose that initially no force is applied to the transducer and no voltage appears at its terminals. Now suppose that a step input of force is applied. In other words, imagine that a force is applied in a time short enough to be considered negligible and then held at a steady value.

As a result of this applied force, a charge appears across the transducer plates. Let us call that charge Q_0 and the corresponding voltage V_0, where $V_0 = Q_0/C$. The charge is generated when the step of force is applied. While the force is held steady, no more charge is generated.

Will the voltage across the transducer also remain constant?

No. To measure the voltage, current (a flow of charge) is taken by the measuring circuit. This reduces the charge across the transducer plates and so the voltage decreases in proportion.

What is the overall response of the transducer?

I can show this by reference to the circuit of Figure 44 but first of all I must make an assumption. This is that the piezoelectric material responds to the force step by generating a corresponding *charge* in negligible time. It is the subsequent flow of this charge in the electrical circuit which determines the step response of the transducer.

Figure 45 represents the situation just after the application of the force step. The transducer now has an output voltage V_0.

The voltage across the capacitor also appears across the resistor, so the current at this moment must be $i = V_0/R$. The current, which is a flow of charge from the capacitor, will lower the capacitor voltage.

As the voltage falls, so the current falls and, as the current falls, the voltage falls at a slower rate. So this is a situation where the slope of a function (in this case dv/dt) is proportional to the value of that function (v) at all times. In other words, this is an *exponential function*. Putting it symbolically, at any time after $t=0$, $q=Cv$.

$$\frac{dq}{dt} = i = C\frac{dv}{dt}.$$

$$v = iR = CR\frac{dv}{dt}$$

or $\quad\dfrac{dv}{dt} \propto v.$

This first-order differential equation is of the form

$$\frac{dv}{dt} = kv = \frac{1}{\tau}v,$$

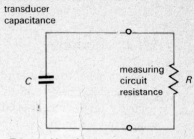

transducer
capacitance

measuring
circuit
resistance

Figure 44 A circuit representing a piezoelectric transducer connected to a measuring circuit

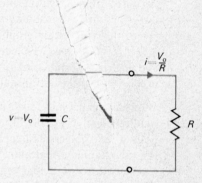

Figure 45 The situation just after a step force is applied

exponential function

where τ is the time constant, equal to CR.

The solution of this first-order differential equation is

$$v = A \exp(-t/\tau)$$

(see 'Mathematics for Instrumentation').

At $t=0$ we know that $v=V_0$ so A must be V_0 and

$$v = V_0 \exp(-t/RC).$$

So this circuit, which represents the piezoelectric transducer and its measuring circuit, is another example of a first-order system.

However, although its response is exponential, it differs from the response of the temperature transducer, explained in Unit 3, in one important respect. Figure 46 compares the response of a temperature transducer to a

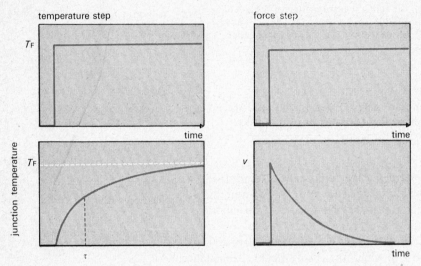

Figure 46   The step responses of a thermocouple and of a piezoelectric transducer compared

temperature step with that of a piezoelectric transducer to a force step. The important difference is that, while the temperature transducer has a delay in its response, the piezoelectric transducer representation of Figure 44 responds immediately to the step but cannot maintain its output subsequently. (The simplified model of Figure 44 ignores the fact that *no* transducer can respond *immediately* to a step input. The delay in response is, however, negligible compared with the subsequent decay time constant $\tau=CR$.)

The form of the step response of the piezoelectric transducer has an important implication: whereas temperature transducers cannot respond adequately to changes occurring in times less than the order of one time constant, piezoelectric transducers cannot respond adequately to force changes occurring in *longer* times than the order of one time constant. In other words, they are not suitable for very slow changes in force. Their output falls if the changes in force become slower and, in particular, is practically zero after a steady force has been applied for more than about five time constants.

5.4   Measuring circuits

Voltage measurement

The discussion up to this point has considered the *voltage* response to an input step. An alternative technique is to measure the charge directly but, before I discuss charge measurement, let us consider some of the features of voltage measurement.

37

You saw, in the last section, that the flow of charge from the transducer into the load resistance caused the output voltage to decay exponentially after the application of a force step.

If the transducer is used to measure a time-varying force, this effect reduces the amplitude of the voltage output at low frequencies.

This happens for frequencies of the order of $1/\tau = 1/RC$. (Remember that RC is a time, so that $1/RC$ is a frequency.) Compared with higher frequencies, the voltage amplitude at a frequency of $1/RC$ is reduced by just over 1 per cent. This sets a lower useful limit for the frequencies with which a piezoelectric force transducer can deal.

If a piezoelectric transducer has a capacitance of 10^{-9} F, what is the minimum amplifier input resistance needed if the voltage amplitude error is to be less than 1 per cent for a frequency of 5 Hz?

We require that $1/RC$ be less than the lowest frequency, that is

$$\frac{1}{RC} < 5 \text{ Hz}$$

or $R \times 10^{-9}$ F $> \frac{1}{5}$ s.

Thus R must be greater than $(10^9/5)$ Ω, that is, 200 MΩ. Special techniques are needed to achieve input resistances of this order.

A second difficulty arises if the amplifier has to be some distance from the transducer. The linking cable will have capacitance of its own and this will effectively be in parallel with the transducer capacitance. Capacitances in parallel add. A cable may typically have a capacitance of 100 pF per metre. The capacitance of piezoelectric transducers is typically from 50 to 2000 pF, so that situations can easily arise in which the effect of cable capacitance is quite appreciable.

SAQ 14

SAQ 14

A piezoelectric force transducer has a charge sensitivity of 10 pC N^{-1} (one pC is 10^{-12} coulomb) and a capacitance of 500 pF. It is used with an amplifier chosen to give an output of 1 V N^{-1}. What gain should the amplifier have:

(a) If it is connected directly to the transducer?

(b) If it is connected to the transducer via 5 m of cable having a capacitance of 80 pF m^{-1}?

You can see from SAQ 14 that changing the length of the cable between the transducer and the amplifier can make a considerable change to the overall sensitivity of the system. But, even if the length of the cable is not altered, its capacitance may change significantly if its shape is made to change or if its temperature changes. These difficulties are avoided by building the amplifier into the transducer casing, but this is not always possible. The transducer may, for instance, have to operate at too high a temperature for the amplifier; or the weight of the transducer may have to be kept as low as possible so as to minimize any changes it makes to the dynamic properties of the object on which force is being measured.

5.5   Charge measurement and charge amplifiers

The difficulties mentioned in the last section are particularly serious for a transducer with relatively low capacitance, in which the effect of the capacitance of the cable is more prominent. This type of transducer is often used with what is normally called a *charge amplifier*. This is not an accurate name, because it is not charge that is amplified. The amplifier acts essentially as a charge-to-voltage transducer, in the sense that its input is charge and its output is voltage. It is an amplifier in the sense that the available electrical output power is greater than the signal input power. It consists of an amplifier with negative feedback provided by a capacitor. The amplifier can be an operational amplifier of the type described in Unit 2 with its non-inverting input connected to ground. To review its operation, try SAQ 15.

charge amplifier

SAQ 15

SAQ 15

The operational amplifier in Figure 47 has its non-inverting input connected to the zero potential (earth) wire, via a resistor to balance out offset current. Two other resistors are connected as shown. Assume that no current flows into the inverting input of the amplifier.

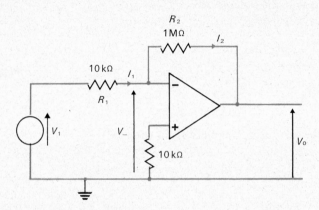

Figure 47   Diagram for SAQ 15

(a) Obtain expressions for the currents I_1 and I_2 in terms of the three voltages V_1, V_- and V_o.

(b) Work out the gain V_o/V_1, assuming that the differential voltage gain of the amplifier is infinite.

(c) Work out the gain V_o/V_1, assuming that the magnitude of the differential voltage gain is 1000.

It would be a good idea to go over sections 8.4 and 8.5 of Unit 2 if you cannot do this self-assessment question.

In order to explain the operation of the charge amplifier, I shall consider two cases, starting with one which is highly simplified and dropping some of the simplifying assumptions in the second. Before doing so I shall recapitulate the relationship between current and voltage in a capacitor.

$$i = C\frac{dv}{dt},\tag{13}$$

that is, the current i flowing through a capacitor is equal to the capacitance C multiplied by the rate of change of the voltage v across the capacitor.

Figure 48 shows the charge amplifier, with its feedback capacitor C_f, connected to a transducer. The transducer is now represented by a charge source shunted by capacitor C (C includes any cable capacitance). The current flowing into the inverting input is assumed to be negligible.

Case 1. Amplifier with infinite gain

The theory is similar to that of SAQ 15, except that the current i_1 is now dq/dt, the rate of flow of charge from the generator, and i_2 is the current in the feedback capacitor C_f.

39

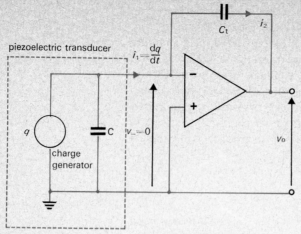

Figure 48   A piezoelectric transducer connected to a charge amplifier

The voltage across C_f is $v_- - v_o$ and, as in the self-assessment question, v_- must be zero for infinite gain, so that the voltage across C_f is $-v_o$. Thus the current I_2 is $-dv_o/dt$. The two currents are equal if no current flows into the inverting input. This gives

$$\frac{dq}{dt} = -C_f\frac{dv_o}{dt},$$

which can be integrated directly to give

$$q = -C_f v_o + \text{constant},$$

where the constant can be made to be zero if the amplifier is set to give zero output voltage for zero charge.

We therefore have

$$v_o = \frac{-q}{C_f}.$$

The output voltage is proportional to the input charge, which is why the amplifier is called a charge amplifier. Notice that because the voltage v_- is zero, there is no current flowing in capacitor C. In fact, the value of C, which includes both the capacitance of the transducer and of the cable connecting it to the amplifier, does not even appear in the expression for v_o. Thus the use of the charge amplifier eliminates the problem of cable capacitance. However, it does not eliminate it completely, as the assumption that the gain of the amplifier is infinite is not realistic.

Case 2. The amplifier has finite voltage gain of magnitude A_V

If A_V is not assumed to be infinite, v_- will not be zero, although if A_V is large (say 10^3–10^5) then v_- will still be much smaller than v_o. In this case some of the charge resulting from the force applied to the transducer will form a current in C as well as in C_f. The relationship between v_o and q can be found in a manner similar to that in Case 1. The result is

$$v_o = \frac{A_V}{C + C_f(1 - A_V)}q.$$

A_V is normally very large compared with 1, so that we can write

$$v_o = \frac{A_V}{C - A_V C_f}q.$$

It is quite easy to choose $A_V C_f$ to be very large compared with C, even if C represents the combined effect of the transducer and the cable capacitance. (For example, A_V could be of the order of 10^5 at low frequencies and C_f could be approximately equal to C, so that $A_V C_f$ would be about $10^5 C$. We then have

$$v_o \approx -\frac{q}{C}$$

as before. This shows that the charge sensitivity v_o/q is effectively independent of the cable length if a charge amplifier is used, which is an advantage over the voltage-measuring amplifier of the last section.

In the case of voltage measurements, discussed in the previous section, the output voltage was $v=qC$. The time constant, and the corresponding frequency response, were determined by the product CR, where C was the transducer capacitance, including cable capacitance. If an amplifier were used to measure the output voltage, R would represent the amplifier's input resistance.

If a charge amplifier is used, however, the output voltage becomes $v_o=-q/C_f$, and the corresponding time constant becomes R_fC_f, which depends only on the properties of the amplifier. This can be determined once and for all, independently of cable length.

One further practical consideration is that the input current to the charge amplifier is not zero. In fact it is a small, but finite current. To provide a path for this current other than the feedback capacitor C_f, a resistor must be included in parallel with C_f, as shown in Figure 49.

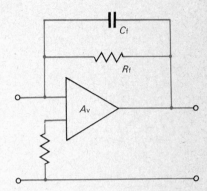

5.6   Temperature effects

Changes in temperature affect piezoelectric transducers in a number of ways. The resistance of the piezoelectric element decreases with increasing temperature.

Figure 49   A charge amplifier with a resistive path in parallel to the feedback capacitor

The natural piezoelectric crystal most commonly used is quartz. It has the advantages of having a low variation of piezoelectric coefficient with temperature and a relatively high sensitivity. Some transducers use various types of ceramic. One factor in choosing a transducer with a particular ceramic is the maximum temperature to which it is likely to be exposed. Thus must be less than the Curie point of the ceramic, as that is the point at which the piezoelectric polarization is lost. The Curie point varies from 120 °C for barium titanate to 570 °C for lead metaniobate, two of the ceramics used.

5.7   Transducer example 6: A quartz force link

Figure 50 includes a photograph of a quartz 'force link' (i.e. load cell) together with a sectional diagram of the force link and its specification. It is intended for the measurement of compressive and tensile forces. You can see from the half-sectional drawing that two quartz discs are used. These are pre-loaded, that is, they are kept in compression, by being squeezed between two nuts screwed onto the pre-loading bolt. The common face of the two quartz discs is electrically connected to the central terminal of the coaxial output socket. The outer faces are both earthed to the metallic structure.

A compression force to be measured is arranged to act along the main axis, and is introduced via bolts screwed into the ends of the transducer. Because the quartz discs are pre-loaded, a tensile force is sensed as a *reduction* in the compressive force applied to the discs and generates a *change* in charge of opposite polarity to that generated by applied compressive forces.

When the force link is used in conjunction with the manufacturer's charge amplifier, the output voltage is independent of cable length and has a maximum value of 10 V. On the most sensitive range 1 V N⁻¹ is obtained.

Most of the items in the specification should be clear to you, but the term 'resonant frequency' needs some explanation. At this frequency the transducer has an increased output. The upper end of the band of frequencies over which the output is independent of frequency is limited to about a third of this. Resonance is explained in Units 8/9/10 in connection with acceleration transducers. Piezoelectric acceleration transducers are described there too.

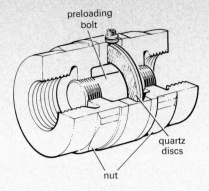

preloading
bolt

quartz
discs

nut

SPECIFICATION

Table 5

Measuring range
±40 kN.

Charge sensitivity
20 pC N⁻¹.

Resonant frequency
37 kHz.

Stiffness
2100 N μm⁻¹.

Overload
10% of full range.

Maximum shear force
15% of full range.

Threshold
< 20 mN.

Linearity
Within ±1% f.s.d.

Error for eccentric loading
±5%.

Insulation resistance
5×10^{13} Ω.

Capacitance
100 pF.

Thermal sensitivity shift
-2×10^{-4}/deg C.

Working temperature range
From −40 °C to +120 °C.

Overall length
72 mm.

Diameter
40 mm.

Weight
485 g.

Figure 50   A Kistler-Swiss type 9351 piezoelectric force link

Section 6

Summary

This unit deals with transducers for force, torque and pressure, and with the displacement transducers which are used, both as secondary transducers within the force, torque and pressure transducers, and as transducers in their own right.

The force, torque and pressure transducers use *elastic sensing elements* to transduce their measurands to strain or displacement. Some basic terms are revised: *mass, force, acceleration; pressure, force; shear stress, shear strain, shear modulus; absolute pressure, gauge pressure, differential pressure; head* and *pressure; torque*.

The sensing elements described for force transducers are the *proof ring* and a *solid cylinder* or *column*. Either of these elements may be used in the proprietary *strain-gauge load cell* whose specification is given.

Sensing elements for torque may be either a proprietary *torque transducer*, or simply an existing shaft subjected to the torque, upon which strain gauges may be mounted.

The longitudinal strain in a strain gauge mounted on the curved surface at 45° to the shaft axis is given by the gauge sensitivity $e/T = 1/(\pi G a^3)$.

For pressure transducers, *diaphragms, diaphragm capsules, bellows* and *Bourdon tubes* are used as elastic elements.

Resistive potential dividers (*pots*) are used as linear displacement transducers. Linearity is affected by electrical loading. *Wire-wound types* have finite resolution. *Carbon-film, conductive-plastic* and *ceramic–metal mix* types have better resolution. A proprietary pressure transducer is described which uses a pot to sense the deflection of a Bourdon tube.

Three types of magnetic displacement transducer are described:

The *differential transformer* produces an a.c. output voltage whose amplitude is proportional to displacement and whose relative phase indicates the direction of displacement. The *phase-sensitive detector* conditions this output signal and produces a d.c. signal of amplitude and polarity proportional to displacement.

The proprietary differential transformer, whose specification is given, is a 'd.c. to d.c.' type with built-in electronics.

In *capacitive transducers* mechanical displacement varies their *capacitance*. The capacitance of a *parallel-plate capacitor* is given by $C = \varepsilon_0 \varepsilon_r A/d$. Displacement varies the area A, the separation d or the relative permittivity ε_r of the *dielectric*.

Capacitance-measuring circuits depend on the relationship between the amplitude of the a.c. current through a capacitor and the a.c. voltage across it: $I = V\omega C$.

The a.c. open-circuit output voltage amplitude of a *capacitor-bridge circuit*, using *differential capacitors* of either the *variable-separation type* or the *variable-area type*, varies linearly with the displacement δ:

$V_0 = \delta V/2d$ (d = separation at balance);
$V_0 = \delta V/2l$ (l = plate overlap at balance).

A proprietary *capacitive differential pressure transducer* is described, which uses a differential variable-separation capacitor to sense the displacement of a diaphragm, either side of which the two pressures act.

Section 1

1.1

1.2
1.3

1.4

1.5

Section 2
2.1
2.2
2.3

Section 3
3.2

3.3

3.4

Section 4
4.2

4.3

4.4

Piezoelectric transducers exploit the *piezoelectric effect*, whereby an applied force generates a proportional charge.

Quartz crystals and *ceramics* exhibit piezoelectricity. Because of their stiffness, they are essentially useful as *force* transducers rather than as displacement transducers. The *step response* of a piezoelectric transducer is *exponential* and decays with time. It is another example of a *first-order system*. It differs from the temperature transducers in that its *low-frequency response is limited*. Measurements of the output *voltage* of a piezoelectric transducer suffer from dependence on *connecting-cable capacitance*. The use of a *charge amplifier* makes the measurement essentially independent of cable capacitance: $v_o \approx -q/C_f$, where q is the generated charge and C_f the amplifier feedback capacitance, provided $A_V C_f \gg C$ (the capacitance of the transducer plus cable). *Temperature* affects the resistance of the piezoelectric element. The *Curie point* must not be exceeded, or the polarization will be lost, in ceramic elements.

A proprietary *quartz force link* is described which has two quartz discs pre-loaded by a bolt. Either compression or tension may be applied to the link, and is sensed as a generated charge of the appropriate polarity.

Section 5

5.1

5.2

5.3

5.4

5.5

5.6

5.7

Appendix Units and conversion Factors for pressure

Starting unit	Pa	dyne/cm^2	bar	lbf/in^2 (p.s.i.)	kgf/cm^2	inH$_2$O	mmHg
pascal	1	10	10^{-5}	1.45×10^{-4}	1.02×10^{-5}	4.02×10^{-3}	7.5×10^{-3}
dyne/cm^2	0.1	1	10^{-6}	1.45×10^{-5}	1.02×10^{-6}	4.02×10^{-4}	7.5×10^{-4}
bar($\approx$ atmospheric pressure)*	10^5	10^6	1	14.5	1.02	4.02×10^2	7.5×10^2
lbf/in^2 (p.s.i.)	6.89×10^3	6.89×10^4	6.89×10^{-2}	1	7.03×10^{-2}	27.67	51.715
kgf/cm^2	9.81×10^4	9.81×10^5	0.981	14.22	1	3.937×10^2	7.356×10^2
inch of water	2.491×10^2	2.491×10^3	2.491×10^{-3}	3.61×10^{-2}	2.54×10^{-3}	1	1.868
millimetre of mercury	1.333×10^2	1.333×10^3	1.333×10^{-3}	0.019 34	1.359×10^{-3}	0.5352	1

*Atmospheric pressure on the earth's surface varies from 900 mbar to about 1010 mbar.

The table should be used as follows:

1 starting unit = conversion factor × unit at head of column

e.g. 1 bar = 10^5 Pa,

   1 kgf/cm^2 = 14.22 lbf/in^2.

Self-assessment answers and comments

SAQ 1

Force=mass×acceleration.

Denoting force by the symbol F, mass by m and acceleration by a, this relationship may be written

$F=m\times a$.

The SI unit of force is the newton, with the symbol N (m kg s^{-2}).

SAQ 2

Pressure=force per unit area.

In cases where the force F is uniformly distributed over an area A, the pressure p is given by

$$p=\frac{F}{A}.$$

The SI unit of pressure has been given the name pascal and symbol Pa (N m^{-2}=m^{-1} kg s^{-2}).

SAQ 3

See Figure 2. The total force causing shear is represented by the force F. It acts over the area A, parallel to it, and *deforms* the body from its unstrained shape (shown by the dashed lines).

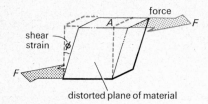

Figure 2 (repeat)   The definition of shear strain

Shear stress=force per unit area *parallel to the force*,

or, denoting the shear stress by S,

$$S=\frac{F}{A}.$$

The SI unit of shear stress is the pascal Pa.

Shear strain is a measure of the change in shape of the body, measured as an angle (in radians). In Figure 2 the angle is φ.

Shear strain=angular distortion of an object from its unstrained shape.

Shear strain is a dimensionless quantity. The SI unit is the radian rad.

Notice that this angular distortion changes the shape of planes of material lying at right angles to the area A (over which the force F acts) and including the line of action of F.

The shear modulus G is the ratio of shear stress to shear strain, for strains sufficiently small that the stress and strain may be considered to be linearly related.

$$\text{Shear modulus}=\frac{\text{shear stress}}{\text{shear strain}},$$

$$G=\frac{S}{\varphi}.$$

The SI unit for shear modulus is the pascal Pa.

SAQ 4

Absolute pressure is measured with respect to the pressure of a perfect vacuum.

Gauge pressure is measured with respect to the pressure of the surrounding atmosphere. So

Gauge pressure=absolute pressure−atmospheric pressure.

Differential pressure is the difference between pressures.

SAQ 5

The *gauge* pressure p_{gauge} exerted by a liquid whose surface is at a height, or head h, above the measuring point is given by

Gauge pressure=density of liquid×acceleration due to gravity×head,

$p_{\text{gauge}}=\rho g h$,

where ρ is the density of the liquid.

SAQ 6

Torque is twisting force. See Figure 3. The torque is defined as the product of the *force* and the *distance* from its line of action to the axis about which the twisting takes place.

The SI unit of torque is the newton metre (N m).

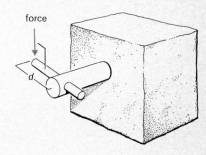

Figure 3 (repeat)   The torque on the shaft is the product of the force and the distance d from the centre line

SAQ 7

The torque on the shaft is 2.5 N×200 cm=0.5 N m. The shear strain produced by this torque is thus 0.5×0.01.

$\varphi=0.5\times0.01=0.005$.

The longitudinal strain e in each gauge is half that, or 0.0025. Gauge 1 is in compression. Gauge 2 is in tension.

SAQ 8

To obtain a linearity better than 1 per cent, $25R/R_{\text{m}}$ must be less than 1.

For $R_{\text{m}}=10$ kΩ, this gives R less than 400 Ω. The next lowest value available is 220 Ω. The maximum sensitivity is obtained with the maximum voltage.

$$V_{\text{s(max)}}=\sqrt{(1.5\text{ W}\times220\ \Omega)}=18.2\text{ V}.$$

Using this voltage the sensitivity is 18.2 V/139 mm=0.131 V mm^{-1}.

SAQ 9

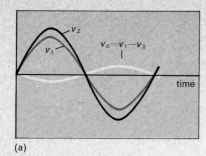

(a)

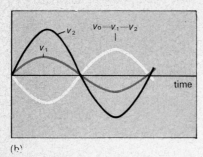

(b)

Figure 51 *The secondary voltage waveforms in the differential-transformer load cell: (a) for smaller compressive forces; (b) for larger compressive forces*

SAQ 10

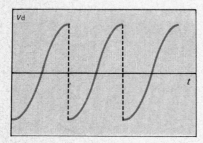

Figure 52 *The output of the phase-sensitive detector when its input is 90° out of phase with the primary voltage. It has an average value of zero*

SAQ 11

The voltage amplitude is proportional to $L\omega I$, so for constant I, it is proportional to ω. Thus the sensitivity is also proportional to ω. Increasing displacement increases the reluctance of the air gap, thus decreasing the inductance of the coil ($L=N^2/R_m$). Thus the voltage decreases with increasing displacement.

SAQ 12

$$C=\frac{\varepsilon_0 A}{d}=\frac{8.85\times 900\times 10^{-6}}{10^{-3}}\text{pF}$$

$$\approx 8\text{ pF}.$$

At a separation 0.5 mm closer, $d=0.5$ mm and $C=2\times 8$ pF$=16$ pF.

Change in capacitance$=+8$ pF.

At a separation 0.5 mm greater, $d=1.5$ mm and

$$C=\frac{2}{3}\times 8\text{ pF}$$

$$\approx 5.3\text{ pF}.$$

48

Change in capacitance ≈ -2.7 pF.

Notice the clear non-linearity in the capacitance–displacement relationship.

SAQ 13

Considering the cylinders as curved parallel plates,

$$A=2\pi\times\text{radius}\times\text{length}$$
$$=2\pi\times 20\text{ mm}\times 300\text{ mm}$$
$$=12\pi\times 10^{-3}\text{ m}^2$$
$$d=0.5\text{ mm}.$$

Maximum $C=\dfrac{\varepsilon_0 A}{d}=\dfrac{8.85\times 12\pi\times 10^{-3}}{0.5\times 10^{-3}}\text{pF}$

$$\approx 670\text{ pF}.$$

Minimum $C=0$.

Total change ≈ 670 pF.

The capacitance–displacement relationship is linear with a constant of proportionality of

$$\frac{670\text{ pF}}{300\text{ mm}}\approx 2.2\text{ pF mm}^{-1}.$$

SAQ 14

(a) Using $q=Cv$, the amplifier input voltage is

$$\frac{10\times 10^{-12}}{500\times 10^{-12}}=\frac{1}{50}\text{V N}^{-1},$$

so that a gain of 50 is required to give an output of 1 V N^{-1}.

(b) The situation is shown in Figure 53. C' is the cable capacitance which is $5\times 80=400$ pF. Since capacitances in parallel add, the total capacitance is $C+C'=900$ pF. The amplifier voltage is now

$$\frac{10\times 10^{-12}}{900\times 10^{-12}}=\frac{1}{90}\text{V N}^{-1}.$$

so that a gain of 90 is now required to give an output of 1 V N^{-1}.

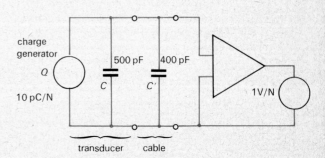

Figure 53 *Answer to SAQ 14(b)*

SAQ 15

(a) The voltage across the 10 kΩ resistor R_1 is V_1-V_-, therefore

$$I_1=\frac{V_1-V_-}{R_1}.$$

Similarly, the voltage across the 1 MΩ resistor R_2 is V_--V_o,

$$I_2=\frac{V_--V_o}{R_2}.$$

(b) The differential gain is $-V_o/V_-$, because the non-inverting input is at 0 volts. If $-V_o/V_-$ can be assumed to be infinite and V_o is finite, V_- can be assumed to be zero, and

$$I_1 = \frac{V_1}{R_1}, \quad I_2 = \frac{-V_o}{R_2}.$$

If the inverting input takes no current, $I_1 = I_2$, that is,

$$\frac{V_1}{R_1} = \frac{-V_o}{R_2}$$

or $\dfrac{V_o}{V_1} = \dfrac{-R_2}{R_1} = -100.$

(c) If the differential gain is 1000, $-V_o/V_- = 1000$, so that $V_- = -V_o/1000$ and, equating currents,

$$\frac{V_1 - V_-}{R_1} = \frac{V_- - V_o}{R_2}$$

$$V_1 = V_- + \frac{R_1}{R_2}(V_- - V_o)$$

$$= \frac{-V_o}{1000} + \frac{1}{100}\left[\frac{-V_o}{1000} - V_o\right]$$

$$= \frac{V_o}{100}\left[-\frac{1}{10} - \frac{1}{1000} - 1\right]$$

$$= \frac{-V_o}{100}1.101,$$

so that the gain $\dfrac{V_o}{V_1} = \dfrac{-100}{1.101} = -90.09.$

Acknowledgements

Grateful acknowledgement is made to the following for material used in these units:

Unit 3

Figure 12: Comark Electronics Ltd; *Figures 14, 15* and *16* and *Transducer 3.2.* Pyrotenax Ltd; *Figures 17, 18, 19, 20* and *21* and *Transducer 3.3:* Rosemount Engineering Co Ltd; *Figures 22, 28* and *29:* ITT Components Group Europe; *Figures 30* and *40* and *Transducer 3.5:* Barnes Instruments Co Ltd; *Figure 36:* Academic Press, Inc. and the authors for R. G. Fleagle and J. A. Businger, *Introduction to Atmospheric Physics*, copyright Academic Press Inc. 1963; *Extracts from British Standards, including BS1904: 1964 re platinum resistance thermometers*, are reproduced by permission of the British Standards Institution, 2 Park Street, London W1A 2BS.

Unit 4

Figures 6 and *7* and *Table 1:* BLH Electronics Inc.; *Figure 21(a)* and *Table 2:* Pye Dynamics Ltd; *Figure 30* and *Table 3:* RDP Electronics Ltd; *Figure 32:* Sybrook Electronics Ltd; *Figure 34* and *Table 4:* SE Laboratories (EMI) Ltd; *Figures 41* and *42* and *Table 5:* Rosemount Engineering Co Ltd; *Figure 50:* Kistler Instruments AG, Winterthur, Switzerland.

Instrumentation

1   Introduction to instrumentation

2   The measurement of strain

3 ⎱ Transducers 1:
4 ⎰ Temperature, displacement, force, torque, pressure

5 ⎫
6 ⎬ Numerical control of machine tools
7 ⎭

8 ⎫
9 ⎬ Transducers 2:
10 ⎭ Acceleration, vibration, velocity, flow

11 ⎫
12 ⎬ Noise in instrumentation systems
13 ⎭

14 Recording

15 Displays

16 Instrumentation in train development